ASTROPHYSICS AND COSMOLOGY FOR NON-MATHEMATICIANS

HITEN SHELAR

Copyright © Hiten Shelar 2024

CONTENTS

Preface

About The Book.

SOME INTERESTING STUFF

Preface

As a kid, I always was interested in knowing about our night sky. One of the facts that surprised me as a kid was that every star we see in the night sky is like our sun. Again, there are 8 planets in the solar system like the ones we are living in. And the fact that we weigh differently on different planets. This is what I learned when I first visited a planetarium as a 5 year old kid. While in the planetarium session, I heard that the time runs differently on different planets. As a kid, I didn't grasp anything about this concept of relative time, but was curious to know about it. One day, when I was around in 9th standard, I was watching Doraemon. It was the most famous cartoon in India back then (Maybe even now). In the episode, there was some depiction of The Theory of relativity, where a man makes a far away spaceflight in the movie in the cartoon and as he returns to earth, he finds his grandson to be of his age. The moment I watched this, my mind picked it up, and I returned to my curiosity which I ever had as a kid. This time I was mature enough to know what was happening. Next, I did not have a time delay to order Einstein's book on

Relativity. Hoping it to be as easy as a 9th standard textbook. As a kid although I didn't have my maths that strong. So I used to read the entire book just reading the theory part. But this didn't impress me since I could hardly learn anything from just reading the theory part and not the math. But as a teenager, I had a burning curiosity to know and learn more. I used the internet to grasp as much knowledge of Relativity. And then later I realized that the subject called "Astrophysics" and "Cosmology" answered all the queries I always asked. While learning basic Astrophysics I got to know about quantum physics in many places, of which nothing I knew back then, and it was required to know it at a certain level for further learning of the subject. I again ordered a book on Quantum Theory, but as 14 year old kid, I couldn't understand the second-order differential equation and many other mathematical tools. Within two years after that, I gathered so much of the non-mathematical knowledge as I could. Until I was just 18 when I had a pretty good grip on maths including tensors, differential equations, etc which one would not like to encounter as an enthusiast for just the theory and the outcome.

Now once I had a solid grip on maths, I gained a lot of knowledge and verified the existing one. And now I realize as an astrophysicist, that so much of the knowledge can be gained non-mathematically after the experience. Well, sometimes I think, what if I time travel in past and give all this non-mathematical astrophysics and cosmology knowledge to the 14 year old me. Well, but past time travel in our universe is not possible. So what I can do is, give this knowledge to the world now. And so what, I have written this book. This book provides that knowledge that everyone enthusiastic about Astrophysics and Cosmology would like to know. and mainly in a non-mathematical way.

About The Book

If you want to acquire knowledge in Astrophysics and Cosmology, you find although tough to deal with complex mathematical equations. This book will serve you this non-mathematical knowledge to an extent. The book is the main juice of non-mathematical knowledge I ever acquired as a part of my astrophysics career. I hope this book serves the best.

We humans have always defined ourselves by by our ability to overcome the impossible.

- Interstellar movie 2014

One

THE SPACETIME

The study of our universe is not just the study of what is beyond the earth or outer space. It's the study of everything. We always believe that something is made up of something and that's how we believe that something exists. We know that every object is made of atoms which are the composition of protons, neutrons, electrons, and so on. And this is how reality makes sense to us. If I tell you to imagine any physical entity to be having no subparts then is that possible for you to imagine? (Try it) No one can do that! It's our very basic approach to figure out that something is always made of something and repeat. So what is our universe made of? For now, let me ask you where you are. And you would give your location. Of course, your address is a name given by a human to a particular place. But have you ever tried to think that where exactly you are in this universe and what are you doing here and what is this universe itself in the first place? Well, lets us keep in mind that we are in space. One should not imagine just outer space now, in physics space means a freedom in dimensions or let's say a length, an area, a volume, or, a hypervolume. Imagine for our ease every point in this space being described by a set of numbers. If you are trying to locate a point on a line, we would just use a number. If you are locating a point on the plane, we would use two numbers to locate the point and three for 3-dimensional things, and so on. We are quite familiar with this math at our schools. Let's now imagine 'time'. Hey, have you ever thought about imagining time? Ok from now we will assume it same as a space coordinate. Yes ! a spatial coordinate. The world around you which you can see is all 3 dimensional and when I ask about your position you give me your location in these 3 dimensions. But let's not forget that for any event we do not describe just the location of the event but also the time of that event. For example, If you not have to miss your train you need to know where that train is approaching and "at what time". So to successfully describe any event we need to treat time as a separate dimension. So let's treat time as a separate dimension, . Let's be clear that wherever you look around the universe you look at the past, even if you are describing anything in space you are describing its past! You are describing an event. The description of this event can only be given completely when you speak out its location in space and time. This space and time is not a separate thing at all as I said one always witnesses and talks about any event in space and its past and so both are not separate at all! So we will not call ' space and time' but rather 'spacetime'. And this is what our universe is made of. The continuum of spacetime is what every constituent of this universe is based on, including the four fundamental forces of our universe. So 'spacetime' (the building blocks of our universe) is made of 3 spatial which we can see and feel and one time dimension that can't be seen but just felt, making it a 4-dimensional entity. So yes, our universe is four-dimensional. This idea of treating time as a separate dimension and this universe was brought by Einstein initially and that has revolutionized astrophysics and scientists' way to look at this universe

Two

THE UNIVERSAL CONSTRAINTS

For the existence of inhabitants of this universe and in fact the universe itself, several universal constraints are made by mother nature. In reality, it is the fact that we as the inhabitants of this universe and the universe as our home have an influence on each other in a way that several rules and constraints are followed. What Constraints? There are many universal constraints in this universe to talk about (which is the part of another chapter), But what one will have in this chapter is the idea of the logic behind their formalism. Some of the examples of universal constraints are the speed of light constancy, shortest achievable length, mass-energy density constraint, etc But why do we have these constraints at the first place? Universe or to be more clear spacetime has a maximum limit of curvature it can exhibit. If the curvature goes too wild say until we get some mathematical infinities over there, then that part of the universe ceases to exist. Infact it would be even rough to be called a part of the universe itself. Provided we have several universal constraints, not every system achieves the mathematical infinities. The reason behind having the universal constraints can be interpreted as the attempt of the universe to peacefully coexist with its inhabitants. Why are scientists afraid of mathematical infinities? Mathematics seems to be describing our universe so effectively to consider anything else. In fact, mathematics is the language of the universe. Physics as a subject has several rules to teach and mixing mathematics helps us to predict the behavior of a particular system and developing the technology. Given the initial information about a soccer ball, one can describe its trajectory with the help of the laws of physics and math. The appearance of the mathematical infinities is not a fault of mathematics but could be due to our formalism of a particular theory to describe the system. If we manage to change or adjust our theory, mathematics would go fine over there.

Strictly speaking, there is nothing special about the speed of light and it's the speed of CAUSALITY which is constant and not the speed of light!' Speed of causality is basically the maximum speed anything could be able to travel at. The value of this constant is approximately 3,00,000 km/s. When one starts accelerating he or she tends towards 'c' and accordingly has its mass exceeded. Correspondingly that particular mass has some spacetime curvature associated with it changes when mass adds up to the system. Relativistic effects like mass, time dilation, length contraction, etc are basically the attempts of nature to make speed of causality as same regardless of any frame. More clearly, The Fabric of spacetime changes itself in a way such that the speed of causality appears the same to every observer in this universe. And this induction of changes in spacetime is what appears to us as the relativistic effects. Like the speed of causality, many universal constants play a role in maintaining this universe along with its inhabitants. Next chapter gives reader a different vision of how universal constraints work.

Three

SPACETIME GRIPS BECOMING THE UNIVERSAL CONSTRAINTS

(A different perspective of universal constraints)

Mass energy tells spacetime how to curve, spacetime curvature tells mass how to move. This grip is what has made the world around us a possible place and this is how a planetary system and two or more mass pair gravity intersection works. This grip but has a limit. Yes, there is a limit to how much stress any mass-mass energy over spacetime. These are not only the limits but are the universal constraints that apply to every system of the universe. What if these universal constraints get violated? Well, then the simple answer would be that the fabric of spacetime breaks down. Has it ever happened? I am talking about black holes. This is the spacetime breakdown caused by the excessive mass-energy density of a star which made it violate the universal constraints of this universe. It's famously known as a 'Black hole'. Another constraint to discuss is the universal speed limit. Yes, there is the fastest speed at which you can go this speed being the speed of light. The closer you go to the speed of light, the more your mass increases and so it requires more and more energy to push the cart to the speed limit. So one requires infinite energy to push the cart to the speed of light and that adds so much mass-energy to the cart that spacetime for it just breaks down.

These are just two of the many universal constraints of the universe which apply to every system in this universe. Throughout this book, we will talk about two reference frames. First is the inertial reference frame, which is the frame where an object has zero velocity. The other is the non-inertial frame where an object has a non-zero velocity. Well, there is nothing truly called a universal reference frame. For example on earth even if we may be at rest but at the galactic scale we are still moving. We use the terms inertial and noninertial reference frame throughout this book to compare two motions in one's locality and not at the extended scale. These universal constraints bring out the relativity in the events. Yes, the parameters like time, length, mass, etc are relative. In special relativity, more the speed one has relative to the fixed inertial frame the more one will disagree on the parameters of time, mass, and length with the fixed inertial frame. Well. Why do we have these constraints in the first place? The simple answer would be that this is how everything in this universe is defined. These universal constraints are just grip for keeping anything in this universe existing in it. Everything thing in this universe has the meaning of its existence deeply associated with the fabric of spacetime. And having a spacetime breakdown in case anything just makes these things non-existing. The spacetime continuum which our universe is entirely made of has a grip over every constituent of it. This grip has made gravity and other fundamental forces a possible thing. These fundamental forces are the basic tools that created our universe in the way it is. So somehow this grip gives everything in this universe the freedom to get into the non-inertial frame of reference (when forces interact and things accelerate) but the same grip is even having a boundary or a limit for how much extent to have this freedom. This extent is what we call the 'Universal Constraints'

Four

THE SPECIAL THEORY OF RELATIVITY

(Introduction)

Let's say there are two frames of reference, inertial (at rest) and noninertial (moving). Let's consider a train where Alice is enjoying his game of throwing a ball upwards and bob who is seeing his friend Alice from the train platform. From Alice's perspective, her ball is going upwards and for bob, his ball is tracing a curvilinear path as the train is moving (As Bob is in the rest frame). The action of throwing that ball looks so different in both the reference frames. One can also refer to Einstein's famous lightning bolt experiment for this. An event perceived by two observers is completely dependent on the spacetime coordinates they posse which is where you are and where you will be at a given time. Let's take another example where Alice and Bob are travelling in two individual cars at speeds v and u (v > u). Now according to Alice, Bob is travelling at speed v-u. But a person at rest on the footpath will still see bob travelling at v. So according to Alice, Bob is travelling at a slower speed than what it seems to be in inertial frame of reference. Now if an inertial observer flashes a beam of light travelling at C (speed of light), what should be its speed with reference to Alice and Bob (think !). The answer is C (It's not a typing mistake). This is the law of constancy of the speed of light in vacuo (vacuum) . Well its not speed of light in generality but the speed of causality (cause and effect) which is constant and has a value of c. From causality I meant that not even the Information can travel at speed more than the speed of light. Our Universe has constrained its participants with this speed limit. It will take an infinite amount of energy to reach this speed for us, which is practically not possible. But , What approach universe make to preserve its own set of rules or constraints made for its residents ? The moment Alice starts accelerating herself she starts getting towards the speed of causality, and as she accelerates more, the world around her (outside the train's window) starts changing itself so that the speed of causality or speed of light remains constant for her. As she achieves a speed closer to the speed of light she will notice shortening of the length of the objects outside her frame (Fitgerald contraction) . One more approach the universe makes to resist anything violating its constraints is the time dilation effect due to which after the journey has completed Alice and the pedestrian (inertial frame) won't agree to the total time passage anymore. More accurately, spacetime will dilate from its initial status of rest. As soon as one starts accelerating, her spacetime coordinates match no more to the one who was initially rest with her. These effects occur for a non-inertial frame as every inhabitant of this universe is constrained by the universe by the universal speed limit (the speed of light or causality) c . Special relativity has a set of transformations (equations) that connect inertial and non-inertial frames which calculate the relativistic effects like length contraction , time dilation , etc . Effects of special relativity are rare at the macroscopic scale and at the atomic and subatomic scale it's a normal thing to observe as it takes comparatively less energy for subatomic particles or particles with less mass to achieve the speed near to the speed of light . The special theory of relativity is a vast field of study in Astrophysics and Quantum Mechanics and we will require this theory In may parts of this book.

Five

THE GENERAL THEORY OF RELATIVITY

(INTRODUCTION)

Newton in his time explained gravity as a line of force acting between two masses A and B . But in reality that is not quite true . His equations although work well in ordinary cases or say in some limit (Newtonian limit) , globally they fail . His equations doesn't explain what gravity is and where it got originated from. Einstein's famous General theory of relativity explains it greatly. According to general relativity, gravity is the curvature in spacetime. What brings this curvature is the energy or mass density. Einstein once had a thought experiment of an upgoing and free-falling lift from which he discovered that gravity is same as the acceleration which was the turning point of relativity. When one starts accelerating, his spacetime coordinates start changing from its initial rest frame and if gravity is the same as acceleration! Can it be interpreted as a change in spacetime coordinates? Of course! It's like if one starts accelerating her spacetime coordinates will change and vice-versa i.e. if one's spacetime coordinates are changed he will start accelerating! That's great! Energy and mass density curves the spacetime (from the flat one) and spacetime curvature gets mass-energy into the acceleration . If Bob has his spacetime coordinates initially flat then as soon as he enters curved spacetime he will feel the acceleration (as space-time is curved), which is the gravity itself. Well! Like a sheet of rubber, spacetime can be bent anyway one likes, all you need is the energy and mass and density at astronomical scales to curve it. Again the General theory of relativity like newtons theory of gravity is not universal like at the center of the black hole and what holds there is the Quantum Mechanics . Mass and energy density has an effect on spacetime by curving it and spacetime has an effect on mass by giving it a direction in its motion . We all are in the influence of the common spacetime curvature (by earth), our spacetime coordinates thus are identical. Everything in this universe traverses a geodesic in spacetime, which may be thought of as the shortest path one can have in spacetime flow. When spacetime is curved time dilates due to which time on Jupiter ticks slower than the earth (as Jupiter due to larger mass deviates spacetime more than the earth). Your flow of spacetime according to your frame of reference will never halt. This is the Principle of Relativity "Law of Physics Will Always Remain The Same In Any Given Frame". Whichever complex spacetime manifold you get into, you won't feel the relativistic effects of spacetime curvature unless you communicate with an inertial frame (outside the spaceship window) Thus it can be said that the speed of light and the laws of physics are not relative. General relativity math is all about calculating the spacetime curvature created by a given mass-energy density, by applying various coordinate transformations.

SIX

THE GRIP OVER SPACETIME AND GRAVITY

Well, for anything to work smoothly, it needs a set of rules. Yes to have a grip over humans we have government. In subsections like in the office, we have several sets of rules an employee should follow and so to have a grip over them, for kids we have schools which again have several sets of rules. Does our universe which is made of spacetime have the same grip over its inhabitant including every atom to the ray of light and so the rules to follow? Yes of course! The rules of society or the rules of humans have nothing to do with spacetime. But the fact is that our universe does possess rules for its inhabitants to follow. We at the local scale not even barely go near to violating these rules. We and every object in this universe and this universe which is entirely made of the spacetime continuum are connected by a grip. I would call this grip 'the universal constraints'. Why constraints? Because as one tries to pressurize this grip, the grip eventually becomes a constraint.

The very first thing spacetime is in direct contact with is the ' Mass-energy'. Yes, no other parameter connects spacetime more directly than this. It can be thought of as an ultimate connection between anything existing in this universe and the universe itself. All other parameters are based on the parameter of 'Mass-energy' and it's what gives meaning to every other parameter in physics. Mass-energy and spacetime both have a grip on each other. Let's have a ball freely moving on this 2-dimensional sheet. Now put the heavier ball anywhere on the rubber sheet and now watch the trajectory of the lighter ball. One can perform this experiment at home. What one can watch through this experiment is how the trajectory of the ball changes when a heavier mass is inserted over a rubber sheet. Now Imagine suppressing the 4-dimensional fabric of spacetime to a plane 2-dimensional rubber sheet and imagine the earth as a heavier ball and the moon to be a lighter one! While in the earth-moon case, Moon won't crash into the earth because of zero friction in space contrary to the rubber sheet ball experiment. What I wanted to explain here from this view is that mass-energy tells spacetime how to curve and spacetime tells mass-enegy how to move.

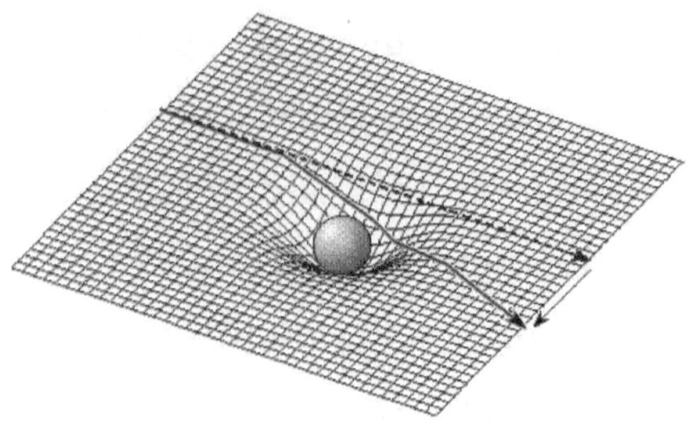

Mass tells spacetime how to curve, spacetime curvature tells mass how to move. spacetime curvature tells mass how to move.

This is the grip I am talking about. Although the rubber sheet example would be a bad idea to illustrate what is going on at a 4-dimensional scale. "MASS-ENERGY TELLS

SPACETIME HOW TO CURVE AND SPACETIME TELLS MASS-ENERGY HOW TO MOVE" I suggest the reader remember the above sentence as it will be needed throughout the book. Well, mass-energy tells spacetime 'HOW MUCH' to curve but not how to curve. So what influences the spacetime to curve in a particular way is the DENSITY OF THAT MASS-ENERGY (Generally speaking) or more specifically 'The Mass-energy density'. Let's treat Mass-energy density as the parameter which tells spacetime "how much and how to curve". The problem is that the density of mass-energy is just one of many factors that tell spacetime how to curve and there are many other factors as well like a frame-dragging effect of the rotating mass which gets some swirling spacetime curvature as it rotates. So in general so many effects act on spacetime curvature and tell it how to curve and each of them has a different nature. To encapsulate all these effects telling spacetime how to curve, we have a "stress-energy tensor quantity" (don't be afraid, I am not going Into detail into the tensors). This quantity covers every mechanism from telling spacetime how much to curve and how to curve considering the multiple effects. From the rubber sheet experiment which was not so accurate, we learned how the moon moves around the earth. But wait damn! We do call it gravity! Yes, spacetime curvature is in fact gravity. Newton was wrong about explaining gravity as a line of force joining the center of the earth and the moon and every other body. It's rather spacetime curvature the stress-energy density created by the earth in the locality of the moon which tells the moon how to move around it. So it's the spacetime curvature, not gravity. So gravity as a "force" loses its meaning. So let's keep in mind that gravity is not a force. The new theory of gravity explained here is the general theory of relativity.

Seven

HAUNTING GRAVITY

(A different Perspective)

We know, if a force is applied to anything, it will move. But we also know that for any object to move we need a force. I mean an object moving without any applied force in real life would be ghostly. Well, the same thing is happening with us and everything around us now. Yes, the objects are moving downward without any applied force and that happens on the entire globe. Isn't that haunting? No. because no one is gonna get afraid of ghosts if the same ghost meets you daily and once you get a habit of it, it is not ghostly at all. If we were born in an imaginary world where gravity never works and then suddenly we made access to gravity, yes then gravity would be so much ghostly. But now in the real world, gravity is our daily life ghost force. This same force has made our universe a possible thing, including our world. We are so much prone to this force. The same thing again for the electromagnetic force, we know two magnets attract but we don't find it ghostly. The rest of the two forces work the same way, but we can't see them interacting at our scale. For a physicist, there is nothing ghostly in gravity. Newton said the force of gravity is a line of force that connect every mass in this universe and gifted the world, the equations to reach outer space. Locally and non-relativistic (speed not comparable to the speed of light), newton's laws work quite well. Einstein's description of gravity is beautiful, it's great, it's my favorite, and it describes gravity so well. According to einstein's general theory of relativity, gravity is the curvature in spacetime. And when the spacetime that defines us get affected by this curvature, we feel some acceleration, this is the acceleration of gravity. Everything that has mass energy possesses a spacetime curvature, including the tiniest subatomic particles. Gravity is the weakest of all the four fundamental forces of the universe. So at our scales, the spacetime curvature is negligible. But provided we have a lot of mass-energy in this universe, this weakest force has the strongest gravity force influencers (black holes!) in this universe. Thanks to Einstein's gravity we could explain the trajectories of the planets which newton's gravity failed. For example the perihelion of mercury. The force of gravity is just not the game of mass but also has a role of energy it possesses. Mass energy just tells spacetime how much to curve. But in reality, we need the stress-energy tensor to describe the spacetime curvature possessed by any mass. Newton's gravity never took stress energy into the account. This is the reason it fails to explain the trajectories of many systems in this universe. But yes at local scales and nonrelativistic speeds, they do work extremely well. The fabric of spacetime is the fabric of reality. It is something on which we have our existence in this universe assigned, we can feel space and time around us (not forgetting the fact that it's 'spacetime' and not 'space and time'). And when Einstein proved that gravity which is so much known to us is the curvature in the fabric of reality, then what theory is more astounding than that? (literally!).

Eight

DON'T UNDERESTIMATE GRAVITY

Gravity being the weakest among all fundamental forces has the highest number of force-exhibiting agents. Particular mass energy always carries spacetime curvature with itself (so the gravity) if not other forces. So one can't find any mass energy that does possess electromagnetic, strong, and weak nuclear force but not gravity (strictly) and no vice versa. Somehow the weakest force of this universe has been gifted with the highest number of ingredients that me this force a possible thing. The force of gravity also has the longest range of than other three fundamental forces. So eventually every mass is going to attract every other (if provided enough time). The abundance of mass-energy in our universe has given rise to gigantic structures like stars, and black holes. Galaxies, superclusters, etc in our universe. Black holes are among my favorite among them. Their size range from the size of an atom to a thousand time larger than our solar system. What makes these guys very famous in astrophysics is their strong gravity from where no light and not even information is supposed to escape. I prefer calling black holes a beast because of their appetite. Their normal snack contains stars that are many times larger than our sun and while doing so they continue to become larger. The most famous black hole is the stellar black hole. The process of their formation would be discussed in the next chapter. Black holes are the strongest gravity influencers in the universe, in fact, no force influencer of whichever force is stronger than the black holes. So the weakest force of the universe has the strongest force influencer in this universe (Let's keep this in mind). The stellar-sized black holes are normal-sized. The gigantic, supermassive, and ultramassive 48 black holes are the ones which are located in the center of the galaxy. Quasars, which are one of the brightest objects in our universe are powered by these black holes. After next two chapters we are going to talk about black holes (I was to continue with gravity , but we will require a knowledge of black holes to understand gravity in some cases, so lets complete the black hole section entirely.)

Black Holes and the Fabric of Reality

A point of no return from which even light can't escape. I am talking about Black Holes. Black holes can be thought of as an Infinite space-time curvature, where at the singularity General relativity fails! We know that light always traverses the straight path, so not does near the black hole as the fabric of reality (Space-Time) is itself curved there, making light traversing a curved path. Light basically is the oscillating Electric and Magnetic Fields in space or photons (packet of energy) that carry some energy and as spacetime is curved it has an effect on that energy flow. Other than this light always traces a straight path. Black holes are misconceptualized as two dimensional holes. In reality black holes are an infinite curvature in four-dimensional space-time, curvature in three dimensional space makes it a spherical hole and a hole in time is not what we can see. The flow of spacetime inside the black hole can be compared to the water flowing into the drainage in the washbasin. What makes the black hole unique from other celestial objects in the universe? The thing is that for every celestial object we can successfully apply the field equations of Einstein to get spacetime curvature, but at the center of a black hole one gets Bizzare solutions that are not acceptable. This point is famously known as the singularity. What I think is, whenever the participant of the universe violates its laws it cuts itself from the universe, for example, we haven't detected any particle yet traveling at ultra-relativistic speed as they have already got themselves out of this universe. Similarly, what I think, is whenever an energy mass density makes up a particular spacetime curvature whose value seems to be violating the limit insisted by the universe it cuts itself from the rest of the universe and this is what we perceive as a black hole. We have all sizes of black holes in our universe ranging from microscopic to ultramassive black holes. Black holes are black and thus difficult to target, they are detected by the effect they have on their background like lensing of light due to spacetime curvature. Nothing is permanent in this universe and so does the black hole. Black holes do evaporate, they emit Hawking radiation which takes their mass out of them.

Let's take one interesting journey of an astronaut Cooper, who decides to get into the black hole to explore it. As cooper nears the event horizon time starts dilating for him until he stops at the event horizon. An inertial observer much far away from the black hole will be tricked as cooper never passed the event horizon, but as the principle of relativity holds cooper according to his own frame has passed the event horizon successfully. So who is real cooper inside the horizon or outside it? (think!) the answer is both the observers are correct (That's crazy!). The events one witnesses are completely relative, they are dependent on the space-time coordinate of one posse. Thus there is no reason here to agree with one of the observers as there is nothing as the benchmark event in this universe , If one gets her spacetime coordinates changed one can even dodge a particular event. Spacetime is the fabric of reality, changing it may result to change in the reality itself for that particular observer or frame of reference.

Nine

THE BEAUTY OF PHYSICS

Physics is beautiful and it successfully explains our universe to a great extent if not entirely. For example, the law of conservation of energy says that the total amount of energy in this universe is constant. And as mass is energy, we have a corresponding law of conservation of mass. Let's say a particular thing after falling inside a black hole loses its existence in this universe. So does that include mass itself? Does the total amount of mass energy of the universe decrease in this case? The answer is no. The same applies to properties of charge and angular momentum as they have their own law conservation. When a mass falls into a black hole, that simply does not just vanish, otherwise, that would be a violation of one of the basic laws of physics. As soon as any object cuts itself from the entire universe at the singularity, its mass gets surmounted over a black hole. Yes, its mass adds up to the mass of the black hole. And same is the case of charge as well. What surprises me is when we put a body with some angular momentum associated with it into a black hole. It just simply adds up to the black hole's spin around its axis. It's beautiful to imagine how the very basic laws of physics are conserved here throughout the process. In the process of violent stellar collapse into a black hole, there is no violation of physics. Except for some topics, physics works extremely well to describe our universe. The equation $E=mc^2$ is one of my favorite equations in physics. In this equation on the left side, we have energy, and another side we have mass. General people mostly misunderstand mass with weight, they perceive mass as a force and that is wrong. Mass is but somewhat a different concept. We define energy as the ability of any system to do work. And mass-energy equivalence describes mass as energy which is the ability of a system to do work. I mean imagine using no fossil fuel, but mass as energy to push the cart. This is the beauty of this equation. It says that a small amount of mass can be converted into an enormous amount of energy. So mass stores so much energy. I mean mass is energy! And this result is entirely derived from the principles of the special theory of relativity. This mass-energy equation is not what has Einstein discovered, it been found in many solutions in physics. But the credit goes to Einstein because he particularly stated that mass is energy and that's why he truly deserves this credit. Well, to move further in understanding the black hole beasts, I need the reader to know quantum mechanics at a very basic level. So let's go with that in the next chapter.

Ten

QUANTUM MECHANICS IN LAYMAN TERMS

Let's start with the word quantum, it suggests one about quantity or something which is discrete . Quantum Mechanics describes all four fundamental forces of this Universe with this discreteness . Quantum Mechanics basically deals with the quantum theory which explains the behaviour of particles at the subatomic level. Well! what if I tell you that an electron can pass the wall without crashing on it? Is that magic? No, it's the quantum physics. At the sub-atomic world, it's very common for such things to happen. In classical physics, everything is well defined, for you to cross a wall without colliding you may have to wait for billions of years. In classical physics the total mechanical energy is given by kinetic energy + potential energy, in the case of subatomic particles we have the Schrodinger equation, it's basically the quantum analog of the classical total energy equation. It gives the values of the most probable potential and kinetic energy. At the quantum level the parameters like position unlike classical physics is a very weird concept for example in quantum physics an electron can simultaneously be at the two positions . Such particles are called to be in the superposition state .

Scientists have proven through various experiments that light is a wave. And by the photoelectric effect, light shows the properties of a particle, how can a wave be a particle? unbelievable! And if a wave is a particle can a normal world particle be a wave? of course! It has experimentally been proven that a moving particle behaves like a wave in some conditions and like a particle in others. In Quantum mechanics we have the de Broglie wavelength for it, which connects the momentum of the particle with its wavelength itself.

A subatomic particle can never be at a definite position or at an absolute inertial frame, as small quantum fluctuations are always present in this universe, this energy never allows subatomic particles to be at absolute rest, thus they always have some wave associated with them. The wavelength of each and every particle has its reach throughout the universe. In quantum physics there are certain rules and conditions for a particle to follow :

1) Heisenberg's Uncertainty Principle: The more precisely you measure the momentum of the particle, the more uncertain you are with its position and vice versa. Mathematically can be stated as (One can skip the mathematics part) :

$$\Delta x \Delta p \geq \frac{\hbar}{2}$$

here, delta x is the uncertainty in position delta p is the uncertainty in momentum

2) Pauli Exclusion Principle: No two electrons can attend the same quantum state. This principle if tried violated may give rise to the external pressure . For example, the electron degeneracy pressure which maintains the white dwarf from collapsing into the neutron star. This pressure is not due to the underlining new fundamental force, but it's the mathematical tool or an approach of the universe to not let the star violate the Pauli exclusion principle. 108 Sending information faster than the speed of light is also possible by the famous quantum physics phenomenon, also known as the quantum entanglement or The spooky action at the distance. scientists have also physically teleported the photon from one place to other by the same principle, imagine same with the normal matter or with human beings. That's incredible! Quantum physics also explains various astrophysics phenomena which classical physics and general relativity cant

explain. It also explains the mechanism behind the four fundamental forces of the universe as being accompanied by force-carrying particles like photons, gravitons, gluons, etc . For Quantum mechanics to be a complete theory we need some experimental evidence. Quantum Mechanics has its approach in fields like physics, chemistry, biology, engineering, artificial intelligence, computer science, and many more. Quantum technology will be a dominant sector in the study, technology, development, and employment in near future, infact it already has.

Eleven

RECIPE OF THE BLACK HOLES

Let's say we have a star that has a mass between two to three solar masses. In every star, the core is always denser than the outer layers of the stars. This core having some energy density always tries to become denser and denser due to the still inner layers of the star and so it always keeps pulling the star layers inwards. But for a star to maintain its status of a star and maintain its size and outer layers, it has the counterbalancing force against the inward pulling gravity. This counterbalancing force is called the hydrostatic equilibrium. The energy from this hydrostatic equilibrium is derived from nowhere but from the strong force. Yes, the same force which is responsible for binding a proton with a neutron. A newborn star begins by fusing hydrogen to helium until there is not enough hydrogen to maintain the star and then helium starts fusing to carbon, after which carbon fuses with helium to form oxygen, oxygen plus helium produces neon, neon plus helium produces magnesium, magnesium plus helium gives silicon, silicon plus helium gives sulfur, sulfur plus helium gives argon, argon plus helium produces calcium, calcium plus helium gives titanium, titanium plus helium gives chromium and at last chromium plus helium produces the iron. Enough! in this battle between the strong force and gravity let's accept that strong force has a limited fuel supply in the battle. Because iron is just unfusible. But the stellar mass is always present to make the star more than ever before. This turns the star into an object called a white dwarf, which maintains its layers by quantum mechanical pressure (will be discussed later) called the 'electron degeneracy pressure'. This pressure has enough potential to counterbalance the inward pull of gravity of the stellar mass for the eternity. And it does not run out of fuel, as it does not run on fuel! The electron degeneracy pressure is not due to any underlining new force. It's just nature's way to avoid the violation of physics. But we are dealing with the recipe of a black hole, so we are ready with enough mass to make degeneracy pressure fail as well! That's the reason why I took a star with a mass of 2 to 3 solar masses heavy as the more the mass, the stronger the spacetime curvature or the pull of gravity is. A star less massive than we assumed would just have ended up at white dwarf. Our sun is expected to have the same fate. Let's get back to our ingredient star with a mass of 2 to 3 times the mass of the sun, having gravity stronger enough to break even the electron degeneracy pressure. The main ingredients of the white dwarf which provide an electron degeneracy pressure are mainly basically electron protons and neutrons. Having our pull of gravity enough, the electron degeneracy pressure fails and the electron and proton in the white dwarf are forced to combine to form a neutron and a neutrino particle. And then we have a thing called a 'Neutron star' with an upcoming BOOM! These booms are the most violent events ever recorded. This boom is called a supernova. This boom is caused by shading the upper layers of stars very violently. But what causes that bump? This bump is caused by a ghost particle that travels like a ghost without interacting with any mass. I am talking about a Neutrino particle. This particle is continuously thrown out of the sun and trillions of them pass through us and everything around us like a ghost. every single second. These particles interact with a weak force (which is stronger than gravity!). But keeping in mind that the outer layer of the star is so dense, there remains a good chance of very few out of a very very high number of neutrinos to interact with the stellar upper layers and blow it in form of a supernova. Which is the most violent event in the universe. So the most violent event in this universe is caused by the least interactive particle in this universe. Isn't that beautiful? Back to the recipe of the black hole again. So now we have a neutron star with us followed by a boom which we just had a long discussion on. The neutron star again has its quantum degeneracy pressure called the neutron degeneracy pressure (will be discussed later).

Again it is stronger enough to keep the neutron star stable till the eternity. But as I said we are prepared with mass high enough (2 to 3 solar masses) to still break throughout this quantum mechanical pressure aswell. Now after this we have our beast, a black hole! A place inside which every law of physics fails including the general relativity, black hole is a breakthrough through spacetime. Our laws of physics simply won't work out as their basis which is completely assigned to the existence of the fabric of spacetime, which just does not exist inside a black hole.

Twelve

A STAR AND A BLACK HOLE

In the last chapter, we discussed that simply having mass-energy is not going to create a black hole instantaneously! Somehow the star could turn into a black hole without just even forming a star, what we needed was just pure mass-energy. But in reality, the majority of mass-carrying particles in this universe do possess other properties as well and so that fundamental forces other than gravity always come into account. These same forces forces forces maintain the star layers in our ingredient star before it turns into a black hole. In absence of these three fundamental forces, gravity would just have formed black holes, and nothing else. But the star and a later black hole formed from it although having the same mass are very different from   each other. Black holes' gravity is much more violent. So much violence that the fabric of spacetime breakdown in its center. So what makes a star a 'star' and a black hole a 'black hole' is the mass-energy distribution. A black hole is very much denser than a star. Both have the same mass, but what keeps a star's density less is its hydrostatic equilibrium for which the strong is responsible. So it's not that we have bulk atoms and we get them together and instantaneously we have a black hole. We have to wait for the stellar fuel to end and again cross two degeneracy pressure barriers to break the fabric of spacetime and form a black hole. But eventually, a star 2 to 3 times massive is going to form a black hole. And this time between a star to black hole transition is what gives chance for life to arise on earth, giving birth to our world. The fusion energy of a star, derived from the strong force which has no direct relation with spacetime, whatsoever is maintaining the spacetime curvature of a star from breaking into a black hole. This is amazing to know how it's a strong force whichh has no direct connection with fabric of spacetime but is still controlling spacetime indirectly although temporarily.

Thirteen

VIOLENT AND NONVIOLENT BLACK HOLES

Black holes are the strongest gravity influencers in the universe. These beasts can be further classified into violent and non-violent ones. The brutality of the black holes speaks a lot of past about the stars which they had formed from. Except when they collide with other black holes which kind of mess up with those imprints. These imprints are left during the star-to-black hole transformation process, which is very violent. Let's say a star with 10 solar masses will transform into a black hole faster than a star that has a mass of 4 solar masses. So we can say that the transformation process was violent in the case of massive stars and this brutality gets imprinted on the fabric of spacetime and the fabric of spacetime gets more violently dragged, creating a less continuous flow of spacetime into the center of the black hole. The more continuous the flow of spacetime is, the less violent the black hole. So in the case of stellar black holes, black holes formed from a massive star is always more violent than a less massive one. This applies to only this case although. In the case of supermassive black holes which sit at the center of many galaxies including ours, the spacetime curvature is less violent and we have a continuous flow of spacetime to the point where spacetime ceases to exist.

Brutality in the case of spacetime curvature means how strongly a curved spacetime curvature departs from a flat one. The stronger this happens the more violent the black hole is. Massive stars once done with the fusion process attain transformation into a black hole faster due to huge mass and so the more gravitational pull. The faster this happens, the stronger the flat spacetime around the star departs from flat to in moving curved spacetime and so the more violent it is. If the star is rotating around its axis the then same is imprinted on the spacetime in form of a frame- dragging effect. In the next chapter, we will see how it is to voyage into violent and non-violent black holes.

Fourteen

FALLING INTO A BLACK HOLE

Let's start our journey to what is inside the black hole with no time delay! Currently, we are in the accretion disk of the black hole paving our way inside it. As we get nearer to the event horizon, time gets considerably dilated for us, which ultimately stops at the event horizon. An outside observer (inertial frame) will observe you as someone who never crossed the horizon, even if you in reality have had done that. a way before. As soon as we near the black hole one will see that before falling behind the horizon, some black background obscures our view of the universe around, This is due to the curved path traced by the light near the horizon. To elaborate, light always travels a straight path in reality but provided that the spacetime metric is itself curved which governs the dynamic evolution of light itself makes the light to curve. But the observer who shares the same spacetime curvature as light finds to difference in the trajectory of the light (These are relative concepts). As we near the horizon, the light rays which were traveling straight for some time before we neared the horizon, now take the curved path, and all the background universe be like projecting themselves in form of a particular circular patch in space dynamically enveloped by the black background as we near the horizon. When we almost get over the event horizon the background universe projects itself as a point until it's all gone when we touch the Schwarzschild radius i.e the radius of the event horizon. This is a great view! when we were in an inertial frame we looked at the black hole as a small dark circle in the surrounding universe and as soon as now we were near the horizon it presents itself as a small circular universe in the surrounding

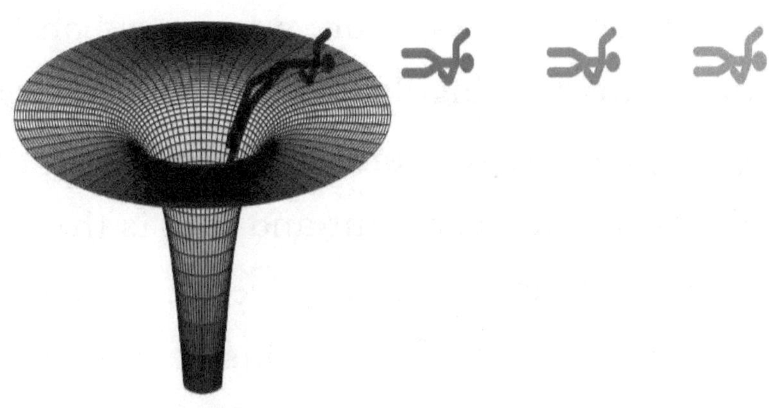

Let's continue our way inside the horizon and now we don't care what the external observer sees! Spacetime switches its roles after one crosses the event horizon, here is where time is what can be perceived as space

and space as a time. Time being perceived as space, one can see past, present, and future as we see the world around us. Well, unless and until we have a bulk being to make a tesseract for us inside the black hole. (as it was done in the interstellar movie) we may enjoy this strange world where time and space had their roles switched! Otherwise, we have to make our way to the singularity. Adding some points before we proceed on our journey to the singularity. The total charge, angular momentum, and mass-energy in this universe remain constant and that is the very fundamental law of this universe. Seemingly every particle which goes inside the black hole has its information lost. But to keep the fundamental laws of our universe unviolated, black holes keep all the quantities of angular momentum, mass energy, and charge surmounted over it. To elucidate, let's take a particle with charge Q, mass M, and angular momentum L. let's have this particle inside the black hole and the black hole gains the same mass as the particle, the same charge, and even the angular momentum gets added! That's so cool! Thus having somebody or something with some angular acceleration with it when falls into the black hole, according to the conservation of angular momentum then black holes will add this angular acceleration to itself by rotating faster just to keep the law of conservation unviolated.

Fifteen

JOURNEY TOWARD THE SPACETIME SINGULARITY

Before continuing our journey to the singularity of the black hole, let's understand what a black hole exactly is. The problem we have in describing the black hole is its singularity part. It's the infinite spacetime curvature, because we don't know what infinity is! according to Quantum mechanics, nothing can be smaller than the Planck length. Classical and general relativity just fails at a Planck scale. In fact, singularity is the point where spacetime loses its meaning. Wait! The universe is spacetime and no universe means what? It's unimaginable! It's like the pre-bigbang condition where time and space had no meaning. Let's still continue further, on our journey to the singularity and as soon as we get there our spacetime coordinates lost their meaning, but still according to the principle of relativity would laws of physics still hold in our own local frame? If No then it's unimaginable to think beyond this! If yes, then there is something for us to experience. Something but what? we have to know the evolution of spacetime after that and as there is no concept of spacetime at the singularity, we can't predict the physics beyond. Thus we may have to physically go there and check that. If singularity is of the planck dimension, what we have at say near to this scale is the quantum foam. This is the region or scale of the spacetime where everything runs in terms of a probability for example a universe may puff out into existence and simultaneously close itself and many more. Its also speculated that our universe started with a random fluctuation over this quantum foam which again collapsed into the supermassive black hole singularity and which in fact we call the cosmological singularity from which our universe began. Thus if we reach singularity then we may encounter the quantum foam which looks similar to like this:

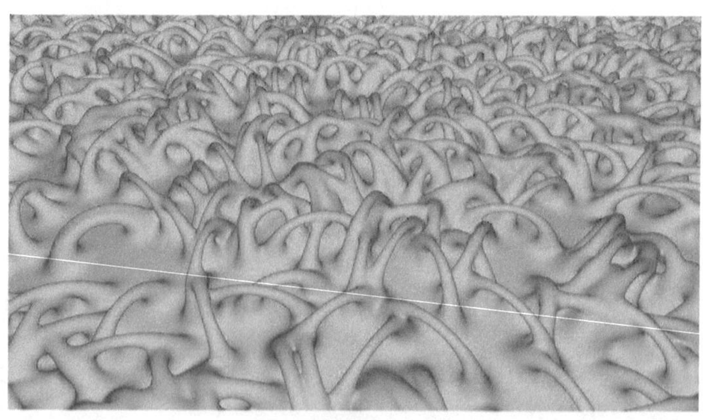

Falling Into a black hole (A Different Perspective)

Well, what's gonna happen depends on which black hole you choose to fall into, violent or non-violent. Let's go with violent black holes first. Let's say you are voyaging near this black hole and you find your way to its accretion disk and are revolving around it at very high speeds. As time passes by you will go near and near the black hole. Let me let you know that you have limited time to boost your rocket off, to escape from the black hole. As you near a black hole, the escape velocity for you keeps on increasing until it attains the value of the speed of light on the event horizon. When this happens, escaping a black hole would be like violating the laws of physics. But again let's not forget that this black hole Is a violent one, so the voyage is not going to be as smooth at all. In the case of violent black holes, the spacetime curvature departs very strongly from the nearby flat universe. So strongly that, your head and legs go in two different Lorentz frames (if you are heading toward a black hole vertically). And this makes you feel a stretching tidal gravity. Yes, the same tidal force by which the moon gets sea waves on earth and by which the moon is tidally locked with the earth. In the case of the earth the and moon, this tidal gravity is only apparent if two points of the body suffering tidal gravity are separated by a distance high enough. This tidal gravity is due to differences in spacetime curvature at points of the body separated high enough. Due to this, the nearmost part of the moon is more under the influence of the earth than the other, the combination of this effect and the centrifugal force on the moon ultimately then has the stretching effect over it, making its one side getting tidally locked with the earth. Near enough to our violent black hole, we have high centrifugal force over us and stretching tidal gravity every micrometer. This tidal force will separate every atom of your body before you make it into the horizon. This process is called spaghettification. So voyaging into a violent would not be a great idea at all! Let's now voyage around a supermassive nonviolent black hole. Again, you have no chance to escape this black hole if you reach the horizon where the escape velocity is the speed of light, and escaping becomes equivalent to violating the laws of physics. In the case of nonviolent black holes, curved spacetime departs from the surrounding universe's flat one smoothly (at least much smoother than the violent black holes). So there is strong tidal gravity but no spaghettification and you can continue your voyage in your spacesuit still further. Once you cross the event horizon you start heading towards singularity. When you reach the singularity, the universe will deduct your mass from you! And then your body has no existence. But why? Because there is no spacetime after that! We and every physical thing In our universe have its meaning deeply assigned to the fabric of spacetime and no spacetime would just mean no existence. So unless there are bulk beings to save you like in interstellar by blocking you inside a tesseract, the end is fixed inside a black hole or not? Well, that would be the part discussed later in this book

Sixteen

ARE BLACK HOLES IMMORTAL?

Black holes have extremely long livability but these beasts like everything else are not permanent in this universe. They lose their mass through a process called Hawking Radiation. This universe is lined everywhere with a thing called 'Quantum Foam'. Just imagine, where there is spacetime there is quantum foam. It's located all over the universe. It is a very basic layer of energy that is located at every point in the universe.

At this level, something creates itself from nothing and again meets nothing. Got jumbled up? Let's go into detail about that. At the scale of quantum foam, some new process happens which cant be seen at our scale. This process is what defines the quantum foam, which is a very basic energy layer of this universe. Imagine a hypothetical zero energy condition of the universe, where there is no quantum foam. Now imagine you have two particles, one normal matter and other antimatter (it's anti-part) in this zero energy (empty vacuum) with the same mass and energy, just with an opposite charge. And after a very short interval of time, these virtual particles return to annihilating themselves under the influence of there own equal and opposite electromagnetic force. That sounds bizarre. I mean how can we have something come out of nothing like magic? And mainly, isn't it a violation of the conservation of mass law? Well, no! Although this law is violated, it happens just for a very very small intervals of time permissible by the heisenberg's uncertainty principle.. Another question is that, can we see normal objects like chairs, spaceships, or anything showing the same behavior? Before answering this question I want to introduce the reader that predicted the existence of the virtual particles in the first place.

$$\Delta t \Delta E \geq \frac{\hbar}{2}$$

This is Heisenberg's uncertainty principle equation which we studied previously in the chapter on Quantum Mechanics. Here 't' stands for the time, 'E' stands for the energy and 'h' is Planck's constant (6.62607015 × 10-34 m2 kg / s). Don't get baffled, I will explain what this equation is trying to convey. The equation says that it's ok to see small energy changes for very small intervals of time permissible by the equation itself. This is nothing but how laws of thermodynathrmics work at the quantum level.

For a scale larger than the subatomic scale, we have a change in energy very high. So this can happen only for a very very very short amount of time. So, can we have a spaceship behaving like virtual particles? The answer is still no. That's because there is a limit to how small a particular time interval be. This is what we call the 'Planck time'. When you calculate the energy associated with the normal day objects, it will come out to be a very large number (at the scale of 10^16 joules). If one substitutes this energy change value in the uncertainty principle, then the amount of time permissible for this object to come into existence comes out to be lesser than the 'planckPlanck, which is the smallest possible time interval possible in this universe. That's why everyday objects can't show such behavior. Coming to the title of this chapter, imagine the same virtual particle pair creation process happening extremely near to event horizon of a black hole. At a point near to the black hole, one has a significant tidal force at every atomic scale distance (the same tidal force which results in spaghettification). This tidal force makes one of the two virtual particles fall inside a black hole and the other escape. The other particle escapes due to a strong stretching effect of the tidal gravity, making a particle nearer to the black hole fall inwards and another thrown out. This also creates a thing called negative mass which we shall discuss later in this book. The thrown-out particle flies at relativistic speeds and escapes the black hole. The kinetic energy which boosted particles at relativistic speeds got from nowhere but from the mass of the black hole itself! How? well, let's discuss that in the next chapter.

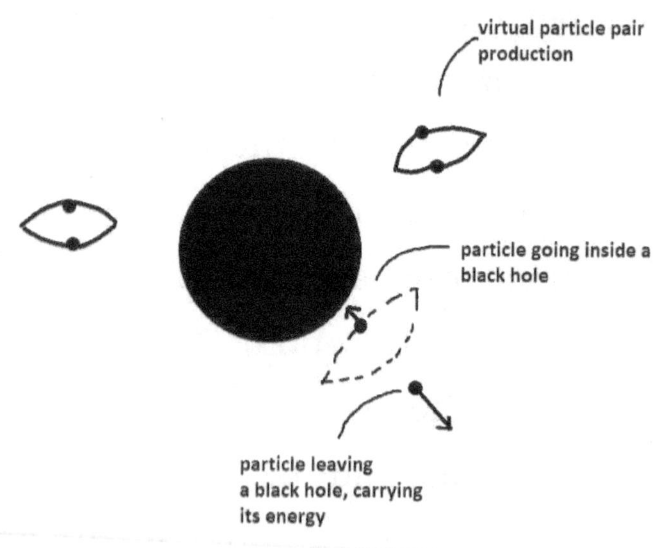

This mass deduction process is very very slow to make the black hole completely get evaporated. By this process alone it might take a time comparable to 10^100 years (one googol!) which is a very very very long period! Normally black holes eat a lot more than what they lost via hawking radiation, so hawking radiation almost has no change for black holes sitting at the center of the galaxy. This universe again proves that nothing is permanent in this universe, This is how beautifully, this universe is designed.

Seventeen

THE BEAUTY OF BLACK HOLES

Humans have started studying astronomy by looking up at the sky and gazing the stars. Back then we used to find something unique in the sky and then search for it with our telescopes and then study it. After frequent studies, we got to know how this system works (at least a basic algorithm). Laterwards we got tools like never before to study this universe. And after that instead of just looking up, we also started to predict the other possible interstellar objects. A black hole was one of these predictions. We predicted every part of these objects so well that we didn't miss every detail of it. And now we have a real image of a black hole :

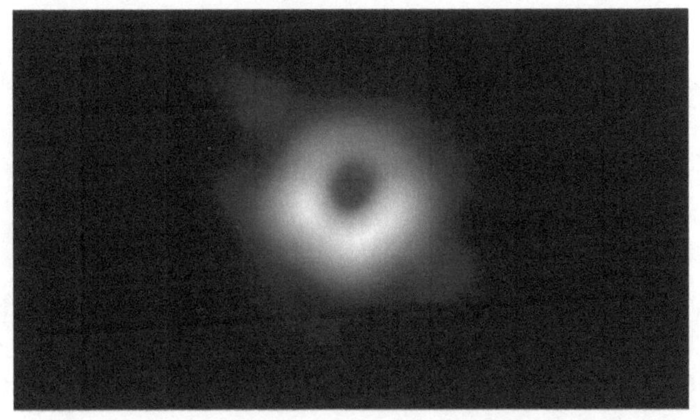

The above image is of the M87 Black hole, which belongs to the M87 galaxy. And it's the first real image of a black hole that was released in 2019 and it still astounds me whenever I look at it. The next image shows the image of a Sagittarius A* black hole which belongs to our Milkyway.

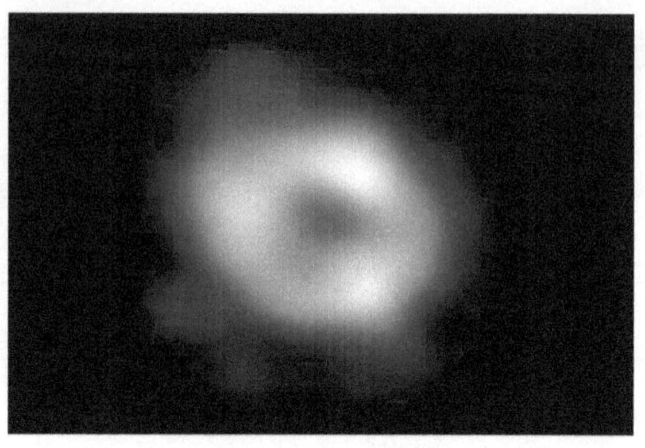

What this image represents is the breakthrough through the fabric of reality. And yes, it is real! We, humans, have successfully predicted something that our naked eyes can't see and finally proved our predictions right by taking a picture of a black hole.

ARE WE INSIDE A BLACK HOLE?

Our universe had its origin from the point called the singularity. The same thing is what we find inside the black hole . But, we don't know how shall we describe it. Quantum mechanics says that quantum foam is what dominates at the singularity. Mathematically singularity is infinitely small in terms of dimension, but we don't know what infinity is

(although quantum mechanics says that Planck length is the smallest achievable length). The only difference between the cosmic and black hole singularity is that the black hole singularity is attracting, while the big bang singularity is the expelling one. Still, there is no problem as we have a time reverse form of the black hole which we call a White Hole. Although we have never seen a white hole, its existence is theoretically possible in our universe. Now when we talk about white holes they also expel spacetime like big bang. But many scientists don't believe in them becausee they violate the second law of thermodynamics. Are we inside a black hole (time reversed) ? Our universe is four-dimensional in nature and its shape is well described by a system of numbers which one assigns say at any point to get a curvature at that particular point, we call this system of numbers collectively as the spacetime metric.

This metric gives us an idea of the possible shape of our universe. If one analyzes the evolution of spacetime inside a black hole or its time-reversed version (a white hole) then it seems very wild as compared to the spacetime evolution in the process of BigBang (when one compare). Currently, our universe is very accurately described (locally) by the FLRW metric (Friedmann–Lemaître–Robertson–Walker metric). Let's say our universe is inside a Schwarzschild black hole (a black hole with zero charge and angular momentum). Although the FLRW metric is different from the Schwarzschild metric, one can find a metric similar or even the same as the FLRW metric inside the black hole's Schwarzschild metric, but only locally ! . Then It will be a really tough task for an observer inside the black hole's (time-reversed) FLRW subdomain to distinguish the metric from the regular universe if it was not inside the black hole. Even now we cant claim that we are not in the FLRW metric subdomain of the larger Schwarzschild metric inside a black hole . Well, the spacetime evolution in our universe seems to be very smooth according to the cosmic microwave background (CMB) and in the case of the black hole, it's comparatively violent. Although we approximated a local FLRW model in the Schwarzschild model of a time-reversed black hole in which we might exist, that approximation was local ! . At the global scale say from the data from the CMB the observation does not commute with the black hole metric. Thus probably we should not believe in that we are inside the black hole. So probably we might not be residing inside the black hole (time-reversed)

Nineteen

SPACETIME INSIDE A BLACK HOLE.

Spacetime flows inside a black hole like water in the washbasin, moving towards the spacetime breakoff (singularity). When you cross the event horizon of the black hole, time and space will switch their roles. And our spacetime becomes time-space'. Under the influence oftime-spacee, you would move in space just in one direction like time and there is no other option. That's because space has become time-like, and as 'proper time' never stops flowing. Inside a black hole, you can't resist the flow of your one-dimensional space. Now, what about the time? The time inside a black hole becomes space-like. So, now here time shows spacelike behavior, and one starts looking at time as one can do in space in the normal world. So one always has the freedom to look at the past, present, and future. Imagine watching various stages of your life as at a same time (astounding!) . However one might like this 'time-space', one can enjoy it only for a short period. This is because of the continuous flow of time-space inside a singularity, where time and space have a breakoff and the concept of time ceases to exist. After the singularity, the spacetime is not special at all. The time In spacetime stops flowing at the spacetime and generalizes Itself to the surrounding hyperspace or extra dimensions.

Twenty

CREATING A BLACK HOLE LIKE HORIZON.

What makes a black hole separate from this universe is the fact that no event inside a black hole can be seen from the outside universe and no event from the outside universe can be seen from inside the black hole. The boundary from which this cut-off of the events occurs in the case of a black hole is called the event horizon. This topic shares an idea that can let us create a horizon like condition around ourselves, if not an event horizon of a black hole. So as one knows that a horizon does the work of separating events, the idea is to create a virtual horizon around yourself that would separate you from that event. Take into the picture, the spacetime diagram on which we plot our velocity curves. Now this time we are extending this spacetime to the entire universe. I know that would be very huge, . So to fit the entire universe under our imagination (at a definite extent) we would do some coordinate transformations. The coordinate under our imagination now has one dimension of space (represented as x-axis) and one time (represented as y-axis) : Now, let's draw a distance-time graph of light on this coordinate plane :

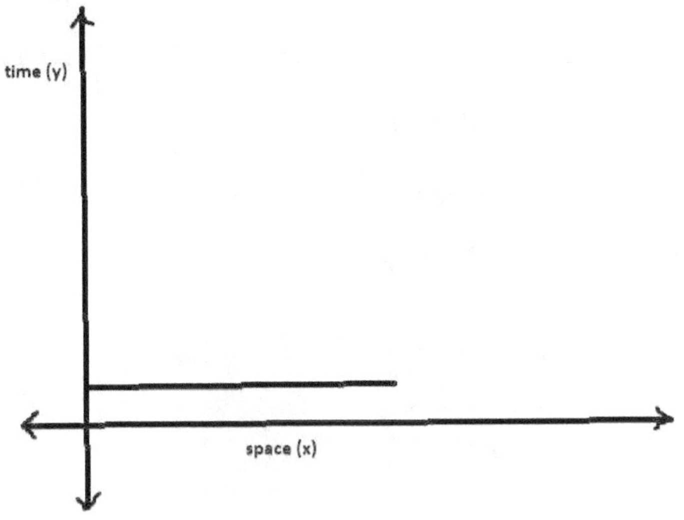

Since time does not elapse for light, we describe the distance-time graph for light as a line parallel to the space axis. Now, imagine our universe as a sheet of paper , now by tying it up on four sides we are avoiding imagining the infinities. Now we would bring the coordinates transformation in the above spacetime diagram in such a way that the graph of light represented above now tilts by an angle of 45 degrees with respect to given space and time axis. Believe, this is a huge curvature in the fabric of spacetime. This curvature help us to bring down huge distances of our universe under our imagination. This transformation gives us the below spacetime diagram as compared to the previous spacetime diagram.

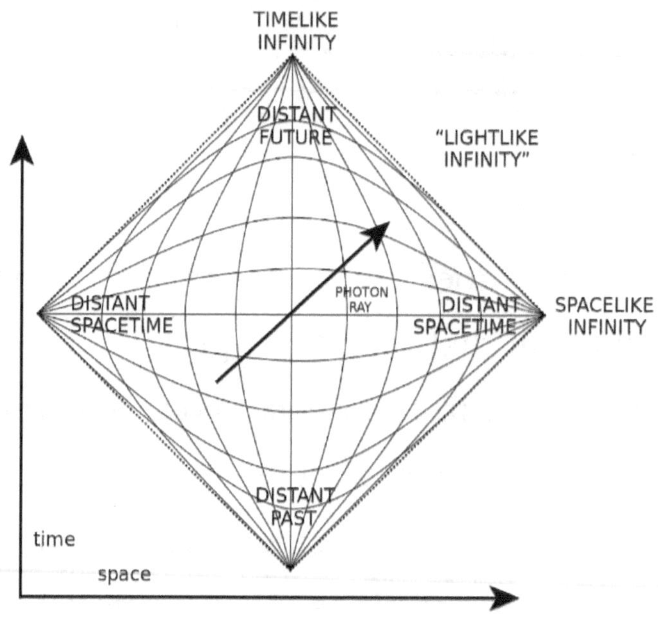

The diagram above is called the penrose diagram. So applying our observations of the universe to this diagram, since nothing in this universe can travel faster than the speed of light, In our new coordinate transformed diagram equivalently, our world line (our trajectory in spacetime diagram) can never bend more than 45 degrees. As that would be the same as moving faster than or at the speed of light. This makes it difficult to bend our worldline as the observers even by one degree . Now lets rotate the 45 degree graph of light around the axis which passes through it at any given point and is parallel to time axis. This gives us a cone whose sides are inclined by 45 degrees in our 4-dimensional spacetime. This cone is called the 'lightcone' since its boundary marks the path traced by a ray of light in our Penrose diagram.

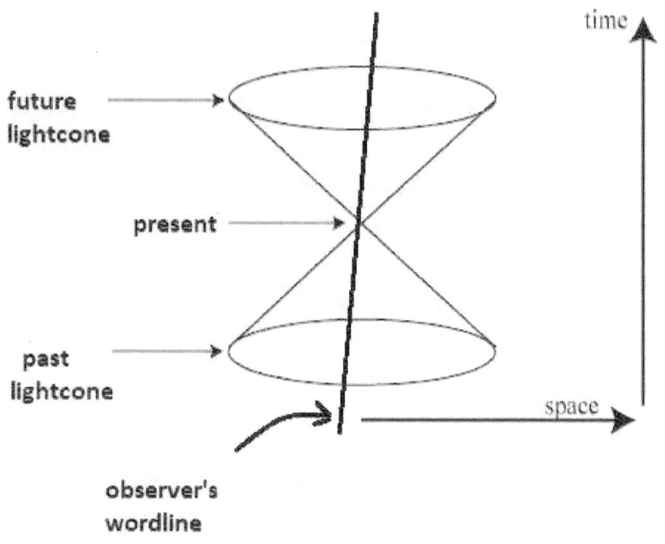

This lightcone is the diagrammatic representation of your Lorentz frame, which you carry wherever you go. Now imagine a star, 4 light years from you goes into a supernova. This information reaches you in the form of light after four years and then you know that a supernova has ever occurred. For any event that happened in past and to be witnessed by you, you would need its information to lie inside your past lightcone. Now, imagine you accelerate in some arbitrary direction, this changes the trajectory of your worldline, depending on how you accelerate. Along with your world line, your lightcone also gets tilted (as it's your Lorentz frame). If you keep this acceleration until the time the light from the supernova reaches you, the light from the supernova (which is moving at 45 degrees) just cannot make Into yout past lightcone and for you to ever observe it (as it's been tilted due to acceleration). So that even remain unevidenced by you. This is just the one event I am talking about, in reality, there would be so many events that are not reaching you, due to your tilted lightcone surfaces. So by accelerating, you are tilting your lightcone surfaces and by doing that you are making many events in this universe 100 remain uncommunicated with you. This is equivalent to creating a horizon around yourself. This horizon is called the 'Rindler Horizon'. We require an ex-extreme amount of energy to boost our rockets to bring a significant tilt in our lightcone. But yes, even a slight tilt can make starlight unreachable to you. So as soon as you stop accelerating and come back to rest or an inertial frame, don't get shocked by looking at the events you never evidenced before.

Twenty-one

BLACK HOLE'S IMPACT ON TIME.

Remember, a black hole is just not a hole in space, it's a hole in spacetime. So it has so many things to do with time as well. Time is the component of spacetime that makes spacetime flow. When one is rest in space, time completely moves on, and when one travels at the speed of light or very near to it, time elapsing stops. This stop in the passage of time also occurs when one nears a black hole. As many might imagine, a black hole is a hole in spacetime from which nothing can escape, not even light, and so on.This has now became an old story. It's now a time to talk about something that a black hole influences and we can't see it but feel it. I am talking about black holes' influence on time. It would be nuts to talk just about time and no spacetime as both (space and time) are attached to each other and it is what our universe is. So let's see what a black hole does to the fabric of spacetime and we shall see in this chapter, how this effect on spacetime can be felt when we feel time. To be honest, I kind of don't believe in astrology or future predictions. But when we talk about black holes, these holes in the fabric of reality do govern our future. Again, I know what I am talking about. The spacetime does flow. We feel the gravitational acceleration. that is because spacetime tends to flow. Even if we might rest in space, the spacetime that defines us does flow. This moving effect of spacetime is due to flowing time. It's a universal law that spacetime tends to flow.

For now, I want the reader to suppress 2 dimensions from three dimensional space, for our ease to imagine. And now we just have one dimension of space and one time in our imagination. Imagine this two dimensional spacetime flowing into a black hole the same way the water makes it into the washbasin. In the spacetime diagram under our imagination, we are somewhere in the middle, which we assume to represent our present. What below our present is the past and above it is the future. The future condition of our spacetime is what going to be our present. For example, if a future worldline has some curvature due to some other massive planet, then that would somewhen reach us in the form of gravitational acceleration. And after that, it goes into the past. This trajectory in spacetime is called a worldline (which in reality is four-dimensional). Every mass energy in this universe has a worldline associated with it. A black hole is a hole in four-dimensional spacetime and it not only swallows the space but also time! Imagine our world line flowing into a black hole like a river (as spacetime always flows). The part first enters is the front side which is the future part (just because time runs from past to future in our universe). Once our worldline goes into a black hole, it has no chance to escape it as could and according to the laws of physics, nothing can. Assuming your future worldline is flowing into a black hole. Your present has to follow the same footsteps automatically due to the flowing nature of spacetime. In other words, the track of the future world line can change by boosting off our rockets. But in the conditon like you are inside a black hole this just wont workout as the escape velocity at the event horizon is the speed of light. Boosting our rocket more than this speed is technically impossible in our universe. So once a sufficient or enough part of your future wordline or future lightcone enters into a black hole, at a certain point you have no governance over your future trajectory in spacetime. At such a point black hole is said to be governing your future. And as once your future world line is inside a black hole, your present is destined to go inside it. That means a black hole governs the future. Do not relate this chapter to your life goals, please. This entire chapter was just about the fabric of spacetime. And remember our destiny is in our own hands, and we are the only driver of it.

Twenty-two

THE BLACK HOLE INFORMATION PARADOX.

The quantities like mass, charge, and angular momentum are the universal conserved quantities. When they get inside the black hole, they get a cut off from the rest of the universe, but not their mass, charge, and angular momentum as per the conservation law. In simple terms, you get some rotating body having a particular mass and charge associated with it and drop it into the black hole and what happens is these fundamental quantities pile up over the black hole, increasing its rate of spin, mass, and charge. The black hole its mass from the things it swallows and also loses the same mass via the Hawking Radiation. The hawking radiation is mostly the photons, which have no information about the object dropped into the black hole. This makes us (In General) question our belief about the conservation of information. This is the Black Hole Information Paradox. Given the initial state of a system we judge its past by the imprints it has on itself as well as on the surroundings and provided there is some information loss, we are judging the system wrong! In fact, the existence of that object is itself an information. The black holes do not even leave that. Ok! take an object remove from it, its massenergy, charge, angular momentum, and now what is left? No existence! Everything is made of mass energy 'at least ' in our universe, it's something which even lies at the tiniest scales of spacetime. In fact, it's the rule of our universe that anyone or anything residing inside it should possess some mass or energy. Mass and energy are the basic fundamental quantity that everything in this universe has. If I tell you to remove mass energy out of any object. A stupid will probably try to burn it but understand you are just breaking the chemical bonds inside it. An intelligent student will say blasting a nuclear mass in a nuclear bomb and you have a part of mass being converted into energy, that feels nuts (Man read the question again!). But a genius will tell to throw that particular object out of the universe. Yes! That's what our vacuum cleaner (black hole) does, it sucks not only the matter, and light but also the mass, charge, and angular momentum and keeps it along with it as per the conservation of mass, charge, and angular momentum. So next time if you want to go out of our universe or existence (equivalently), then one has to visit a black hole, which will get from you your mass, net charge, and angular momentum, and thus now you are ready to go out of the existence. Once an object is gone out of its existence, then so does its information! Once you are out of the universe then the normal physics and its constraints, and laws just fail over there to describe the unimaginable, unexplored reality over there.

Above was the one point of view on the Information paradox. But few theories suggest that the Information In our universe is also a conserved quantity. This os the reason why we are still working to figure It out about whether a black hole Is really leaking the Information. Many physicists firmly believe that Information has to be conserved and so they have theorized various models supporting this. One saying that Information Is leaked by the outflowing Hawking radiation by some way. And other saying that black hole leak by a set of microscoping wormholes tunneling the information inside to the outside universe. Thinking on two perspectives mentioned above, nothing can really be said until we find some evidence of black hole leaking any information.

Twenty-three

BIZARRE OBJECTS IN SPACE

Under our observation at the local scale, it seems that this universe has an abundance of planets, stars, asteroids, and comets. Nowadays we are so much familiar with them that only astronomy-crazy people look up at the sky to wonder about them. But there is something out there still more bizarre. For example, a white dwarf. We imagine this object as a normal white star-like-looking body. But the fact is that its stranger than that. Stranger because it's not made of material things like any of the stars, planets, asteroids, or anything else. It not even can be called to be made of atoms. The same applies to neutron stars as well. They are just simply made up of neutrons. They are so dense that a spoonful of neutron star matter would weigh as heavier as mount Everest. Have you ever heard about quark star? It's a hypothetical star that is suspected to exist in between the neutron star to black hole transformation. Imagine an astronomical body completely made out of quarks. What about the black holes? These objects cant thought to be real at all. That is because the fabric of spacetime on which reality is dependent is broken inside the center of a black hole. So black hole is not a planetary or any stellar object. Although it behaves like other astronomical objects, finally it proves that is not what it seems. Have you ever wondered how planetary strikes would be different from the collision between a black hole and a planetary object? We have many craters which give us a glimpse of how asteroid strikes. But the collision of a black hole with our planet just has a different story. Black holes are mostly famous for their ability to swallow things. Its collision with some other planetary or stellar object is what no one talks about, that is because the planet would be spaghettified the moments before it makes contact with a black hole. But that happens in the case of stellar or massive black holes which have a huge domain that extends far outside the event horizon. So in our example, we are having a black hole that is roughly equal to the size of a hydrogen atom, so it has no larger domains to shred every atom of the earth. We will assume here that the black hole in our example already has its accretion disk built already. The accretion disk of this black hole would be immensely hot. This accretion disk is enough luminous to shine like 100 Hiroshima explosions at a time. That's so bright for an atom-sized black hole. But the same luminosity is even important for a black hole to maintain its accretion disk. Yes, the accretion disk of a black hole is maintained because of the radiation pressure of the accretion disk. This radiation pressure counter-balances the black hole's inward pull, maintaining the stable orbit of matter around it and so forming a stable and hot accretion disk. This accretion disk is responsible for the violent collision of a black hole with our planet. Imagine something brighter like 100 Hiroshimas entering into our atmosphere, heading towards the land. When it reaches very near to the surface of the earth the radiation pressure of the accretion disk will firstly dig the surface of the earth. The matter from the dug part adds up to the accretion disk of a black hole. Once that happens, the black hole swallows the part of the preexisting accretion disk to welcome the new material, I mean that's for keeping the accretion disk stable and due to the reason it gets bigger as it eats. Every time a black hole digs the surface throughout the process it keeps swallowing old material to welcome the new one. This makes a black hole not to stop in between like an asteroid to create and crater and that's it. Rather black holes kind of penetrate through the entire planet creating a global earthquake. So the asteroid collision just ends up creating a crater on the planet, but in the case of a black hole, we have a very deep hole than a wide crater. That is how a planetary collision is different from a collision of a planetary object with a back hole.

Twenty-four

HOW CAN GRAVITY ESCAPE THE BLACK HOLE

Nothing in this universe can come out of a black hole once felt inside it, not even the information. So how is the gravity able to escape the black hole and reach you? Do you think that the gravity you are feeling now is coming out of the center of the earth? The gravity you are feeling right now has nothing to do with the center of the earth. Gravity or spacetime curvature shapes itself based on mass energy and its distribution. The gravity or spacetime curvature, one now feeling is the measure of spacetime curvature in one's locality. The gravity one feels is due to the spacetime curvature one has in his Lorentz frame. So gravity does not travel from the center of the earth towards you.

Does the spacetime curvature at the center affect the spacetime curvature in your Lorentz frame? This is wrong question! Because when we talk about spacetime curvature, no part of it has to do anything with other. I mean we just talk about a resultant spacetime curvature or condition formed by a given mass-energy, its motion and other parameters. So no Lorentz frame in spacetime curvature has to do anything with the other. For example, at the center of a black hole, spacetime ceases to exist, but the surrounding spacetime curvature around the event horizon has to do nothing with that. In the case of a stellar black hole, the spacetime curvature is just an imprint of the star it ever was. So to answer our main question, gravity is just a measure of condition of spacetime one has in his locality. This is a correct way to imagine how gravitational influence works.

Twenty-five

THE FLOW OF TIME AND GRAVITY

Let's say you are near a black hole, your future has yet not been governed by the horizon completely. Have you ever thought about why we are moving toward a black hole? Even if you know that you are at rest. Of course, it's a gravity. But what is pulling you into it at the first place? It's the time! The nature of time in spacetime is to flow. And every mass-energy in this universe has Its role In form of worldline In this spacetime.

When the future part of the worldline flows into some event you have a possibility to meet that event. If your worldline finds itself in the black hole singularity, that possibility turns into definiteness and you inevitably move towards that event which has now become an event in your time than just space. And worldline flows due to flowing spacetime since it is spacetime's part only. And what keeps spacetime flowing is time. So, there is no event of gravitational acceleration until your worldline flows through it and that requires 'time' of spacetime. Whenever we look at the clocks, we just think about whether we would get late for our meetings, school or something else, but we never think of time as a dimension that gives rise to gravitational acceleration.

Twenty-six

THE LENS MADE OUT OF GRAVITY

Every mass has a spacetime curvature associated with it and when light bends by some angle we say that the light is being lensed by the lens of gravity, I mean a gravitational lens. This technique is what Arthur Eddington had used to verify general relativity. This experiment was conducted during a solar eclipse for background starlight to get visible . In a nutshell, the experiment was to calculate the apparent shift in the position of stars when they are accompanied by the sun's gravitational lens. The result was that the stars had their position shifted from the normal. Newtonian gravity could never have predicted it in any way .In gravitational lensing there is not only the spacetime curvature that affects the path of light but also the framedragging effect. Frame dragging effect is basically a swirling drag in spacetime that results from the rotation of the planet or any star or a black hole. Due to the combined effect of the spacetime curvature and the frame-dragging effect, one may get to see the interesting patterns of galaxies and stars whose light encounters this lensing. An interstellar object, say which is beyond some star or planet, due to which it's normally not visible to the observer, may also start getting visible when it gets into gravitational lensing as given below:

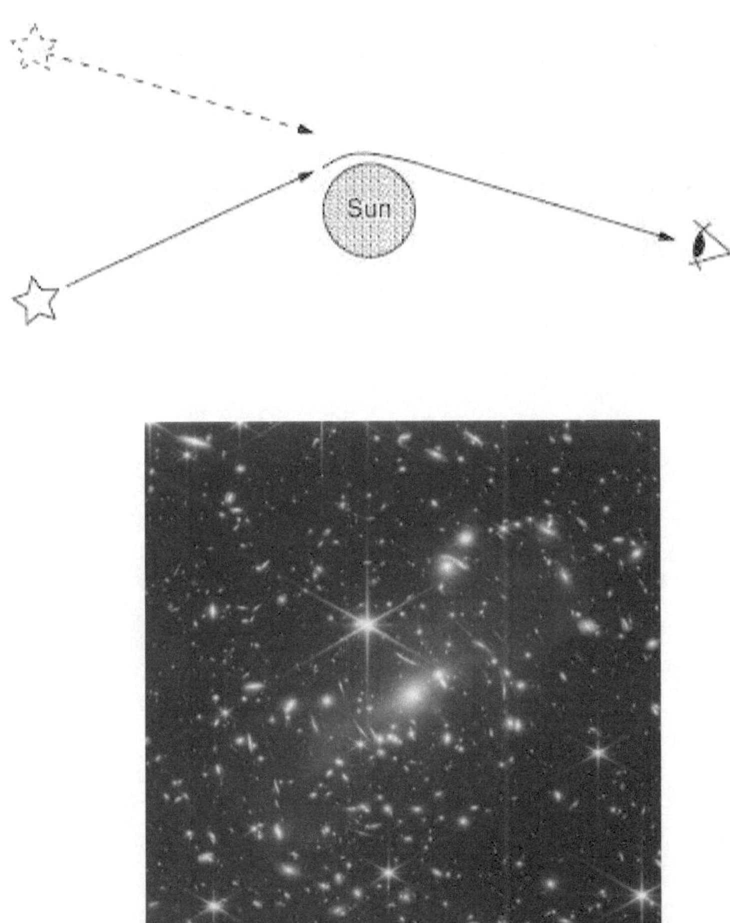

Image by the james webb telescope Here, the distorted image of the galaxy is due to the lensing effect by some interstellar object. As

Black Holes are black, making them hard to identify from the surrounding dark universe, but the effect of lensing they have on a surrounding light of a star or a galaxy makes them detectable. Gravitational lenses can also be used as the agents of the telescopes to see distant objects.

Twenty-seven

WIGGLES IN THE FABRIC OF SPACETIME

From the start, I have made the reader imagine the 4-dimensional fabric of spacetime as a rubber sheet. We can get wiggles in the sheet of rubber, so can we in the fabric of spacetime? Yes! These waves are what we call Gravitational waves. And yes, these are the waves through the fabric of reality. These waves are created by cataclysmic events like the merging of two neutron stars, black holes, etc. The merging of these massive bodies isn't just like a head-on collision. As these objects possess extreme spacetime curvatures, the time they get near to collision, they both first revolve around the center of mass of the combined system. This revolving of two massive bodies creates wiggles in the fabric of spacetime. These wiggles or waves then are sent out at the speed of light. These waves can be detected by the setup, we already have built on earth. And we already have detected many gravitational waves coming from all around the universe. These gravitational waves tell a lot of information about the source they have formed from. Want to know how we detect these waves? Whatever gravitational waves pass through it has a stretching effect on that object. That's because one side of the body momentarily stays under the trough and the other at the peak of the gravitational wave, thus producing a tidal force on that object. This same stretch is what we use to detect the gravitational wave. But it's not simple as may one think. These gravitational waves cant be detected at one's home. For that,t we have a setup called 'LIGO' in Louisiana, USA.

At LIGO (The Laser Interferometer GravitationalWave Observatory), we have an L-shaped system. The arms of this L-shaped system are attached with laser light which meets at a particular point and destructively interferes with each other. When a gravitational wave passes through the arms of the L-shaped system, the arms oscillate. When this happens, the laser light fails to destructively interfere. And flashes of light are detected. These gravitational waves get us a lot of data from their source. These gravitational waves even pass through us as well. But, we still don't feel it. But why? That's because gravitational waves are huge and continuous and our local scale just constitutes a flat part of the gravitational wave. So whenever it passes through us, we barely feel any peak or trough of a wave. At the scales of our planet, these waves might produce measurable stress. In short, the larger the body is, the more it is suspected to suffer a significant stretching and squeezing of gravitational wave. Gravitational waves generally that we encounter are smooth and continuous and are pretty safe. But have you ever imagined how dangerous a noncontinuous gravitational wave would be? Such a wave can produce the spaghettification effects like a black hole, wherever it goes. Although, we don't know how such waves can be produced. Gravitational waves aren't forever. They do lose their potential by doing some positive work. Do you wonder where the gravitational waves get their energy from? This energy comes from nowhere but from the spacetime "gravitational update" that stars ever had before the collision. And then those updates are captured by LIGO providing us the Information about how those neutron stars or black holes merged.

GRAVITY AS A MEDIUM TO SEND MASS-ENERGY.

In last chapter we discussed how we measured gravitational waves by a detector system of LIGO. But did you ever wonder where was the energy which was responsible for the oscillating length of arms of LIGO derived from? I mean gravitational waves were just a medium or form from which that energy was used In doing a work of oscillating the arms of ligo. But where was this energy before, I mean in what form was it present before? It's astounding to know but this energy had came from nowhere but from the mass of two colliding neutron stars itself. Since mass is same as energy. Yes, the mass of the neutron stars there came all the way to here In the form of energy in gravitational waves and this energy was then used in stretching the arms of the LIGO detector

Until now many people might be familiar with mass-energy conversion incase of nuclear fusion. But as we discussed, this is a new form of mass-energy conversion. It is still so astounding to know how the mass of a neutron star several light years away is traveling here in the form of energy In gravitational waves to do some work as It encounters with our planet or any other interstellar planet or object.

You might even heard of a gravitational slingshot where a rocket takes the help of some planet's gravitational potential energy to boost its speed. Ever wondered where this extra kinetic energy comes from? This kinetic energy costs the mass-energy of the planet itself! But this change in mass-energy almost costs nothing to the planet. Due to the fact that even a small mass carries a huge amount of energy. So the boost that the rocket got comes with an exchange of mass not more than not even to consider.

The tidal gravity which brings tides on earth works the same way. Although almost the entire energy of ocean tides comes from the rotation of the earth, it costs a bit of moon's mass aswell. Which is again tiny enough to ignore. Jupiter produces a tidal effect on its moon Europa which brings a stretching effect in it and so creating a tension in the planet, causing it to heat. This the only energy source that maintains europa's temperature and allow water there to stay in liquid state (As sunlight can barely do that at that distance). Again, In what form was this heat energy present before? It's again the mass-energy of the jupiter! And there are many more examples where the mass of a Interstellar object is converted into the energy after doing some work. The medium by which this transfer occurs is nothing but the gravity itself.

Twenty-nine

LENSING GRAVITY BY GRAVITATIONAL LENS

Light can be lensed by the optical lens, so can gravity be lensed by the gravitational lens? Yes of course! But in this case, this is 4- dimensional lensing, now let's see its effects. Gravity can be radiated like light and gravitational lenses can do the work of lensing these waves. Imagine a continuous gravitational wave coming towards the earth. And we have a strong gravitational lens in between like a very massive star which can deflect gravity waves and can lens it on a single point. When we lens sunlight with a simple microscope at the point we get an intensified image of the sun. In the case of light, the simple microscope does the work of collecting the light which falls on its surface and then intensifying it at a point or concentrating it. Here the light that falls on the simple microscope constructively interferes with each other (being a wave), forming an intensified image of the sun. Gravity waves being a wave, show wavelike properties of constructive and destructive interference as well. So as soon as these gravitational waves interact with our gravity lens, they get lensed to an area if not a point (say). Anyone who gets in this area would have to face the intensified effects of the gravitational waves and have to face the extreme tidal force. This system of gravity waves along with gravitational lenses may be can be used as a weapon. Let's say in a war with aliens, we can use this system as a weapon to spaghettify the enemy (amusingly), although we don't know how to get the gravitational waves deliberately. The simple microscope projects an electromagnetic version of a sun on a sheet of paper. And a gravitational lens projects a gravitational version of a source at a particular point in space. As though the mass of the source Is Itself present there.

Let's say we take our lens near a black hole, then would that create a virtual black hole with no mass? Yes but with just one gravitational lens its not possible to lens the entire black hole. As even if we take our lens near a black hole, it would be just a lens a small part of it, and not an entire black hole. And then yes, this part can be confined to a point or diverged or just propagate to some area as it is. This would then create the same spacetime condition over there to the point we wish to project it on. But yes, if a gravitational wave constitutes complete information about a black hole, and say we lens a tiny part of it, then yes a spacetime impression of the black hole can be created. Instead of impression, it would be better if I call it the black hole itself, as there is nothing more real than spacetime.

SINGULARITIES WITHOUT A HOLE OF BLACK HOLE

Gravitational radiation or gravitational waves carry a part of the mass of the system they are emitted from. Imagine a hypothetical black hole suddenly starts decreasing its mass in form of gravitational waves until it has no mass. What are we left with then? Spacetime breakdown! JusttJust that! Have we ever seen anything like these In our universe? Not yet! These are called Naked Singularities. Currently, according to our observation, every singularity is being covered by a black hole around naked singularity would just be then an area or point that cuts spacetime and is not covered by a black hole. But why can't see the naked singularities? The answer is very simple, the information inside a black hole does not make it along the surface of our lightcone. 126 But why is every singularity covered by an event horizon? From the spacetime curvature around the singularity to the event horizon of the black hole, the spacetime is drastically curved as compared to the universe in the neighborhood. There is so much spacetime curvature, that it marks a strong horizon boundary outside. And the lightcones outside the horizon just don't match up with the ones inside the horizon which is in the fascination of the horizon). So can an observer inside even horizon see the naked singularity? Yes, It's,s then possible.

Thirty-one

THE GRAVITY TRACTOR

Say an asteroid larger than a football ground is heading towards us , so is that the end of the world? Well, that depends on the factors of what the asteroid is made of, its angle of collision, the mass of the asteroid, and the velocity of the asteroid. If the asteroid is made of various metals then it is more hazardous than the ones which are mostly composed of carbon. And It also depends on the angle at which the asteroid is impacting. If the collision is taking place say at an angle which is less than 90 degrees then the kinetic energy (energy possessed by every moving object) would be wasted in peeling off the surface of the earth and less violent shockwaves would be created and when the angle of collision is 90 degrees then the most of the kinetic energy of the asteroid is transferred in the form of a shock wave and rest in the form of heat when it penetrates the earth surface. When these shock waves make their way to the oceans then tsunamis will get triggered (flushing out a whole city or country or even the entire world). I should describe the role of mass and velocity collectively as the momentum (= Mass x velocity) is what plays the role here. When the asteroid hits the earth its momentum gets transferred to the earth. I guess an impulse (momentum imparted in a very short interval of time) would be the great replacement of term momentum over here. The entire humanity is in danger. The threat of such asteroids always remains (Although less probable but may or will happen). How do we detect the asteroids? Look up in the clear night sky until you watch any falling star and that is how we do it! These falling stars are just meteorites which are just the chunks of the larger asteroid or any comet. Asteroids really don't have their own light so unless the rays of the sun fall over the asteroid and some traces out of it get reflected, we are not able to target them. And to check for its size, shape, and orbit we use radar technology. Once the trajectory is calculated, one can predict the future path of an asteroid.

But how can we shield ourselves from these hell stones (real ones! literally!)? Asteroids may move at velocity as high as 30 km/s and even more! They have violent trajectories and irregular surfaces. These factors make asteroids harder to detect. And the best time to tackle any asteroid is only when it's more than 7.5 million kilometers away from us otherwise they are considered potentially hazardous ones (provided they have enough mass and velocity to create massive destruction) which are even harder to deal with But how exactly to deflect an asteroid off the track? This how: By Nuking it. Yes, in a situation when an asteroid (potentially hazardous one) is calculated to hit the earth, we can nuke it to either move it to the safer track or to destroy it. It depends on the mass of the asteroid that how many nukes should be detonated to move it off the track (according to Newton's third law). But still, the problem is that even if one turns an asteroid into pieces by nuking it, the meteorite or micrometeorite formed due to the destruction of the asteroid may destroy our satellites. Thus there is a huge loss in the economy. But still, human life is important. It's an effective way to destroy or boost the asteroid off the track. Deflecting an asteroid by the light. Yes, light imparts the momentum like a normal traveling object. But light has no mass! But it has energy and as energy is same as mass, light has its own momentum to impart. The pressure generated by the force imparted by the light due to its transfer of momentum is called the Radiational Pressure. An asteroid can be moved to some other track by this way. It's a good way of deflection as all we need is just high-energy light rays to impart great momentum over an asteroid to deflect its path. Deflecting an asteroid by a gravity tractor : Two masses always have the property to attract each other by the means of gravity. Let's imagine an asteroid (potentially hazardous one) moving towards the earth and we try to deflect it by means of some mass by having a gravitational tug between that particular mass and an asteroid. This mass is basically a moving spacecraft with an ionic propellor sent by us which is set to hover near the asteroid. The spacecraft is placed in such a way that its gravitational pull on the asteroid is perpendicular to the plane of the asteroid belt. Spacecraft and an asteroid if considered as a system then a slight change in spacecraft velocity will affect the spacecraft and asteroid system's momentum as a whole, moving it off the asteroid orbit plane. The force due to gravity due to spacecraft may be very less on the asteroid, but over the years the consistent pull can drag it from its orbital plane and thus securing the existence of our planet earth and the humanity as a whole.

Thirty-two

INTERESTING SINGULARITY

Anything that goes into the singularity, ends up being nothing. But, isn't that making the black hole voyage boring? In this chapter, we shall see how can this voyage be made to continue to some other universe than ending up being something at the singularity. The idea is to connect the singularities of two black holes to each other, creating a tunnel between two universes. This bridge is called a wormhole or Einstein Rosen bridge. It looks like this:

This bridge can also be used to join two universes with opposite running times (physics can be the same or very different in other some other universe!).In that case, a black hole from our universe and a white hole from another would join their singularities. In the universe where time runs backwards, the black hole would show the opposite effects than what it shows in our universe. So, in such a universe nothing can enter a black hole, even light can't reach its horizon, and not even information makes it's its inside. We call this object 'A White Hole'. Again, we don't have any evidence for it to exist right now in our universe.

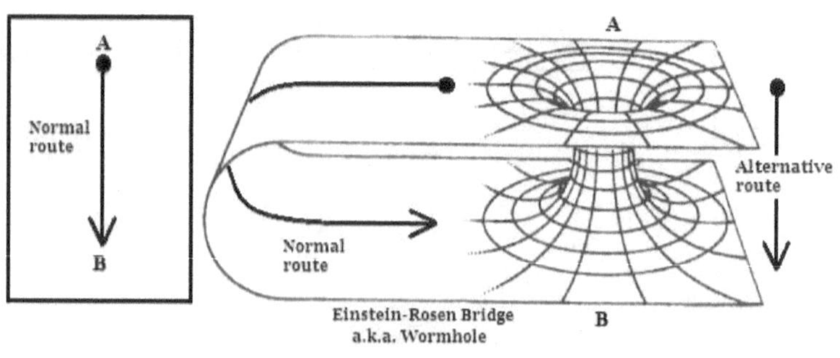

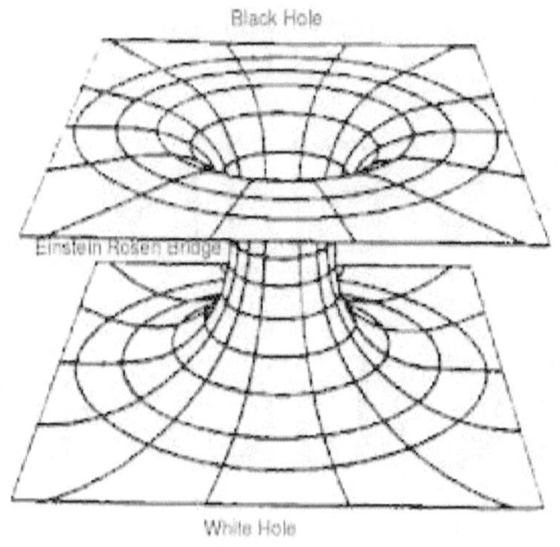

These wormholes can not only join two universes but also two points in our universe alone, creating a shortcut between two points in spacetime. Since a wormhole is a tunnel between two points in 4-dimensional spacetime, it not only can connect two points in space but also time! Creating a time machine for us

Thirty-three

TRAVELING THROUGH TIME

Time is like space, they just have different natures. A wormhole is a bridge between two points in spacetime thus not only can connect two points in 'space' but also 'time' and both of them! Imagine you and your friend Alice decide to do an experiment, where you decide to go in space to have a four year long trip, traveling at a speed comparable to the speed always earth. You and friend Alice decide to meet after four years of your journey.

After the end of the journey when you and your friend Alice meet, You both just wont agree to each other to the total time passage. Originally In your frame of reference even If you took 4 years to do Interstellar travel, the Alice says It took 16 years In her frame. Returning to the world 4 years later from the time you began your space tourism just would be a future time travel by 12 years for you. Yes, when you move at relativistic speeds, time slows down for you, and after you return home you discover that you have time traveled in the future. Now imagine taking one of the mouths of the wormhole to the future by accelerating it at the relativistic speeds by our space shuttle and other at rest on earth . Let's say we go ahead by 12 years in time, taking one of the mouths of the wormhole along with us in the future. Now as the mouth of the wormhole is always connected to the other mouth of it, it can act as a tunnel between which joins a time gap of four years.

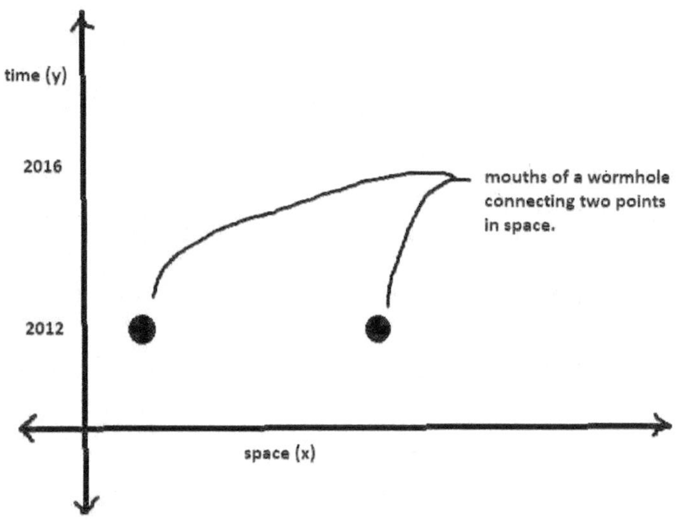

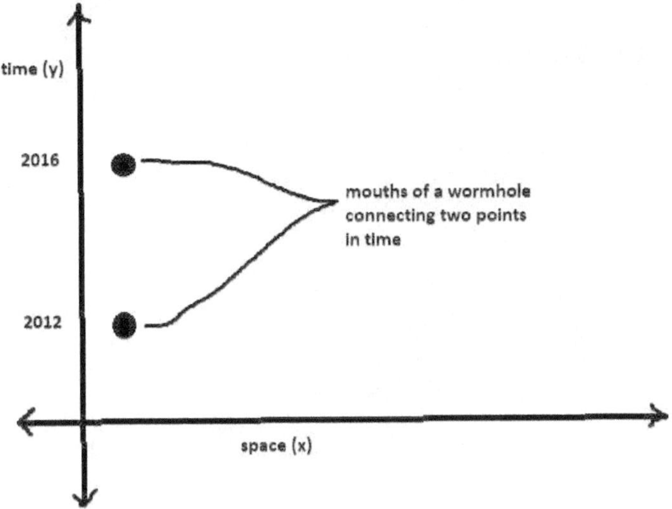

time (y)
2016
2012
space (x)
mouths of a wormhole
connecting two points
in time

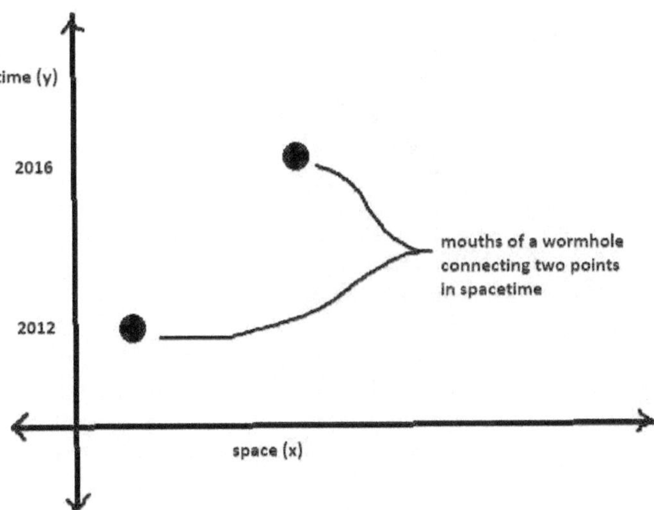

time (y)
2016
2012
space (x)
mouths of a wormhole
connecting two points
in spacetime

So wormholes can connect two points in space, time and spacetime.

Thirty-four

TIME TRAVEL IN THE PAST

One can travel in time in the future along with his Lorentz frame but not in the past. In fact no one can do that in our universe. Our universe or spacetime is made that way so that the time component attached to space always moves in the future. So, it's ok to travel in the future, but not in past (strictly!). It's not even possible by a wormhole. When you accelerate, time slows down for you. Time can slow down until ultimately it stops, which happens if you travel at the speed of light. And if you travel still faster then you travel backward in time.

But, firstly, achieving the speed of light is not possible in the real world and secondly traveling backward in time is still impossible in our universe. Our spacetime is made in a way that time always means a dimension that runs forward and traveling backward in time would simply mean a different spacetim e which is not our universe. So in general we can say that spacetime can never run backward in time. Now, what about that-made wormholes? Incase of wormholes the only people who can go into the future are the people the from past and not the opposite. Because even inside the wormhole the time in spacetime is directed toward the future. So until we have a wormhole made from a white hole and a black hole where spacetime has pulled on either side, time travel in past is not possible. And even If Its possible by a black hole - white hole wormhole, we speculate it to be Inter-universe travel to some other universe where the past we are trying to reach is a present going on there. One of the reasons time travel in the past is not possible is because it creates bizarre time travel paradoxes like the grandfather paradox and many more. Time travel in the past is still a subject of discussion as we find ways to make It happen without the violating of the laws of physics or creating any paradox.

Thirty-five

RECIPE OF A LAB-MADE WORMHOLE

Black hole-black hole and black hole-white hole wormholes aren't complicated to make. Whats complicated is to create a wormhole in the lab. Spacetime curvature is very common in the universe as its carried by everything in this universe. We treat spacetime curvature due to mass as a positive spacetime curvature. All you need to create a wormhole in the lab is the positive and negative spacetime curvature. Positive spacetime curvature is created by a positive nor normal mass. What do you think is needed to create a negative spacetime curvature? A negative mass! Yes, I know that sounds strange. But I will explain, how to create a negative mass-energy density in this chapter. If you are hearing the term negative mass for the first time then please don't confuse it with antimatter (antimatter has a positive valued mass). We are surrounded by an abundance of positive mass around ourselves, the negative might be a very concept for a majority of people. Although we have a possible solution to create it. Let's see how we can do it. We define negative as something lesser than zero. So when we say negative mass-energy density, we mean mass-energy density lesser than zero energy mass density or lesser than nothing! But we know, this universe has a layer of energy attached to it, which we call a quantum foam. Which is the zero-level energy of the universe and not zero-valued energy. This is the lowest energy level attainable in our universe. Which also defines 'flat spacetime' and vice versa. Lesser than this energy is what we define as negative energy!

Take two non-conducting metal plates in a complete vacuum and place them parallel to each other. We know that every mass is accompanied by a wave (quantum mechanics!). All around and in between the metal plates, we have quantum mass-energy in form of these waves and particle state. The more the distance between plates, the more the probability of different energy levels being created between the gap. To make it clear. the quantum mass-energy we are talking about here is nothing but a soup of virtual particles creating and annihilating themselves, the same ones which we discussed In our chapter on Hawking Radiation. When we have a wide distance between the two plates, the energy around a plate and between the plate differs by almost nothing. But as we get the plates closer to each other, they create a limit for a degrees of freedom for the mass-energy waves to occupy the gap between two plates. This happens as only a smaller and lesser variety of mass-energy waves can be built in the small gap. decreasing the variety of the waves that could be formed. The closer we move the plates, the more we put the constraints on the variety of these mass-energy waves that could be formed. At an extremely small gap, only a limited amount of mass-energy can be created inside. Looking at the surrounding now we have two plates surrounded by the zero point energy everywhere around it, except in between them. In between these plates the energy is lesser than the surrounding zero point energy. And what is lesser than a zero point energy? The negative energy! Which is our missing ingredient to create a wormhole. Negative mass repels normal mass and it attracts other negative masses like normal mass. This negative mass would be also useful for us to travel faster than the speed of light (WARP DRIVE SCIENCE FICTION)

Thirty-six

OUR UNIVERSE

Our universe is beautifully designed not only by its constituent but also by the way that every time we try to know about it deeper, it never fails to astound us. The stars we see in the night sky, the air we breathe, the food we eat including our body is made of atoms no one knows which part they emerged from. Life is the creation of mother earth which was nothing more than a dust cloud before our sun was formed. No sun, no planets, meteors just a dust cloud. What brought every particle of this dust cloud together to form our solar system are the four fundamental forces of our universe which are Electromagnetic force, strong force, weak force, and gravity. The fundamental force of gravity is the weakest among these forces. But still, the force of gravity has a major role in making this universe a possible place. It's because it has a longer range than the other three forces. All right, that was part of physics, but what is physics? Is our universe governed by the rules of physics? One should remember that physics is only the language that explains various stuff and phenomenon in this universe, so this universe has nothing to do with physics. Physics as a language to describe the universe has failed many times to explain many things which keep happening in this universe. So physics entirely as a subject is still not complete and we just need a theory of everything which is the center of attention of many scientists in this modern era. In today's busy world, many people just forget that there is something that lies beyond earth. Many people do not even know that there is a subject called astrophysics that studies our universe. Believe me, the more you tend to know about our universe the more you will realize how much you were unknown about the reality you now live in. This new world is no less strange than a science fiction movie. In fact, what makes these things stranger is people's mere ignorance of looking at something outside of the human race. Scientists being human somehow figure this world out and get so much interest in this world which never fails to astound anyone.

Thirty-seven

BIRTH OF OUR UNIVERSE

The birth of our universe is the birth of spacetime. And the entire universe started its life from an infinitely dense point called a singularity. Spacetime ceases to exist at such points. But such points are even comes out to be the birth of the fabric of reality which our universe is. The birth of spacetime is the birth of energy. We describe the big bang as an expansion of the spacetime continuum or our universe from an infinitely small point. After the big bang, there was a sudden increase in the expansion rate of the universe which we call the inflation period. After which our universe is continuously expanding.

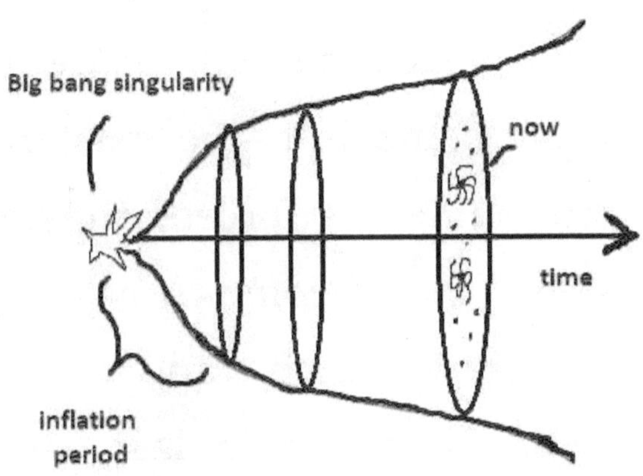

The image above is the 3-dimensional representation of the 4-dimensional big bang. The singularity which gave rise to our universe is called the Big bang singularity or the cosmic singularity, as at that time the entire cosmos was inside it. Many might misunderstand the expansion of the universe as a stretching of the fabric of spacetime, but that's not the case. As our universe is expanding, it's creating new spacetime in itself. So spacetime is not getting stretched, but rather gets regenerated inside. The origin of the universe is the origin of energy. So the origin of energy dates back to the origin of our universe., spacetime curvature or the force of gravity came into the existence. So among all fundamental forces, the force of gravity was the first to come. Next to gravity came the strong force and then the electromagnetic force and in last, the weak force. Between the time period of the big bang and the creation of the force of gravity, all the fundamental forces of the universe were in the form of one super force combined. That's amazing, right? The big bang singularity and the singularity of a black hole are very similar. The difference is that the big bang singularity is the creation of spacetime, while is the end of spacetime. Again, the big bang singularity spews out the spacetime and black hole singularity swallows it down. There is so much similarity in both of them that the opposite of a black hole would be a near-to-perfect replica of the big bang 142 singularity. The opposite of a black hole is a white hole. Wait! Is our universe inside a white hole? Let's go ahead In the next chapter and discuss that.

Thirty-eight

OUR UNIVERSE DURING PLANCK ERA

In total there three fundamental forces in our universe: The Force of Gravity, The Electromagnetic force, The weak force and The strong Force. These forces were not separate as they are now but rather were combined into a single super force. The existence of the super force dates back to the time (Planck time) when our universe was just 10^{-43} seconds old. Before this the only dominating force was the super force, and that's it. After the Planck era (which is the period between the big bang and Planck time), the super force started disintegrating into gravitational, strong, electromagnetic, and weak forces. The gravitational force was the force that disintegrated first after which then was followed by the strong force and after that the electromagnetic and weak force.

Can we get this super force into the laboratory? Well, provided we have the condition hot enough we can do that. All thing you have to do is to get to the Planck temperature (10^{32} Kelvin). If you go hotter than that

that then all the four fundamental forces start unifying into a single super force,

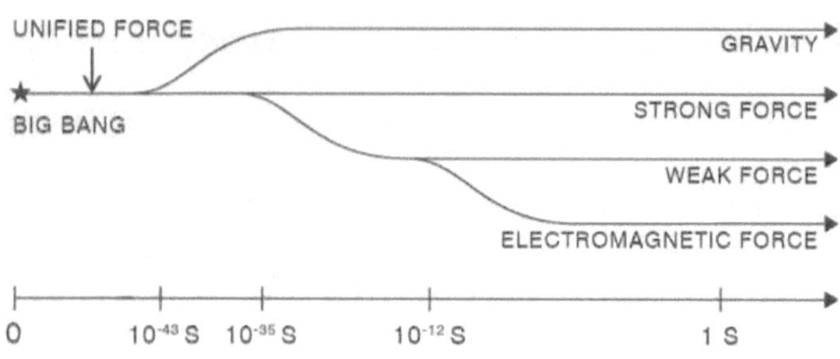

Newton's theory of gravity works well in some if not all the cases to explain our universe and what takes the handle of the whole thing is now the general theory of relativity which still fails at the singularity and thus gives the handle to the quantum gravity over there. Imagine an equation that describes the dynamics of the entire universe and is smaller enough to write on a part of the paper. This universal theory explains everything from the very dawn of creation to the big crunch and everything in our universe. This universal theory is the theory of everything. The Grand Unified Theory is the unification of strong, weak, and electromagnetic forces into a single force not comprising gravity. It's an act of describing the unification of all the fundamental forces of our universe (super force) in physics by the way of mathematics. It will describe the conditions and dynamics of our universe during the Planck era. The theory is still not complete as we have no complete theory of gravity. Except that we have unified the other three fundamental forces. Many of our observations prove Einstein's General Relativity right, except at the quantum scale and the singularity conditions. Shortly, the acceleration of gravity is due to the spacetime curvature according to general relativity and at the quantum scale and the singularity conditions, quantum gravity is what takeover. The unification of general relativity and quantum gravity is the key. Quantum gravity as a theory has to get confirmed via experiments. In LHC (Large Hydron Collider) lets say when we collide two particles at relativistic speeds and say in the soup of new particles created after the disintegration we find a graviton. Then such outcomes would prove quantum gravity true. Up to date, there is no evidence of any graviton found yet. The best theory we know that attempts to unify the Grand Unified theory and the universal theory of gravity which will get us to the Theory Of Everything is the String Theory.

Thirty-nine

THE EXPANSION SPEED OF THE UNIVERSE

Our universe is expanding since the day it formed. Currently, it's expanding at the rate approximately described by the Hubble law.

Which says that farther is the any object from you in space, the more it will speed up from you (Due to cosmic expansion). This law has no exception for the speed of light. So anything that according to the law is moving faster than the speed of light is then moving faster than the maximum speed possible in our universe. Wait, but that's a violation of the laws of physics, as nothing can travel faster than the speed of light. No! physics is still not violated here. The laws of physics are just made for the constituents of the universe and the universe itself can violate them. Yes, it might surprise the reader, but yes it works that way. Even During the cosmic expansion or after the big bang, the universe was expanding at the speed of light and not its constituents. So it's not a violation of the laws of physics. So the speed at which the universe expands can be at any speed, there is no violation of physics here. Does the expansion of the universe have any effect on the spacetime curvature of the earth or any interstellar object? The answer is no! The expansion of the universe is only significant at very wide distances. At a local scale, the force due to expanding universe would be very very less than the counterbalancing electromagnetic and other forces, even gravity. It might just take anything even at the scale of the galaxy, along with it, but will not make it stretch. So, gravity or spacetime curvature or any astronomical object doesn't get stretched due to the expansion of the universe. Our observable universe is roughly 93 billion light years apart in length but provided that the age of our universe is roughly 14 billion years old, we need the part of the universe that far to be traveling more than the speed of light, which is what can be explained by the expanding structure of our universe. Technically, light from the edges of the observable universe should reach us after

41.5 billion light years, but still, it makes its way to us. That's because, even if the object is taken away from us by the expanding universe, but no so the light it emits. So Instead of going along with it, the light gets stretched form the source along with the expansion of our universe. This stretching of light is called redshift as the original light due to this stretching loses its energy. That is why these lights appear in the low energy range to us.

THE BIG CRUNCH

The end of our universe is the end of our universe! So, to find the end of the universe, we should search for an event, rather than seeking for the spatial boundary whichh marks the edge of the universe (as our universe is spacetime and not 'space and time). The universe is going to end in a similar event to the big bang, but just with an opposite functioning. We call this event a Big crunch. So, before any such event happens, we have the continuous and then drastic contraction of our universe. Contraction to what? A singularity! Which would be the same as a black hole singularity, where everything is crushed down to an infinitely dense point. The Big crunch singularity would not bring any stress over the spacetime during its contraction. This is because, spacetime does not really shrink in size, but rather starts to disappear getting smaller and smaller. If we lived in a big crunch era where the universe has just started contracting, we would barely feel this contraction as it's only significant at an extremely large scale. But eventually, as it shrinks further and the temperature of the universe increases, the things we see in the sky with our naked eyes as well as we would start to be affected by it. To escape such a cataclysm there is no choice than traveling to some other universe. Ahead in this book, I will show our progress in this universe and how we always overcome the cosmic threats we might possess in the future. Before we move ahead, lets make the concept of spacetime more clear as it would be needed in the chapters ahead.

HOW TO IMAGINE SPACETIME CORRECTLY

We, humans, love to know new things, we love to explore, and we ask questions about everything. It's our nature to wonder about things and ask questions. We find answers to these questions and get our curiosity satisfied to know something new and bizarre. We are on the planet earth and we have the farthest place any human has gone until now the moon, but our questions have reached the edge of the cosmos beyond it when nobody knows whether the universe has an edge?. It's our habit that we firmly believe in real things or the things which our sensory organs can see, feel or touch. When it comes to spacetime, we do see the space around us and we can feel time. Many people don't even believe time as a real thing, they will simply tell you that time is a clock (which is nuts!).

Even if we believe in space and time, both as real entities, we forget about their connection. Space and time are not separate things, they are connected to each other, forming 'spacetime'. Now even if I say spacetime one might think about space around us and feel the running time. It's hard to visualize spacetime, in fact impossible, as spacetime is four-dimensional and humans cant see 4 dimensions. Now I want a reader to suppress one dimension from the three dimensions of space and imagine spacetime as a dynamic evolution of these two-dimensional space sheets with respect to time. Now since we are looking at 3-dimensional spacetime. It might be easier for a reader to understand it correctly.

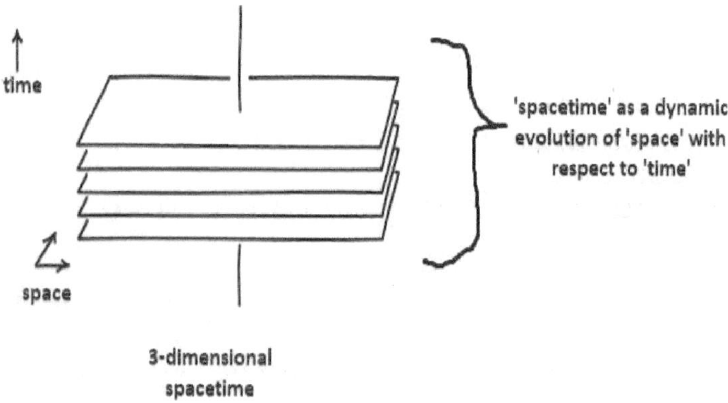

Now feel the real-world 4-dimensional spacetime the same way. Look around the 3-dimensional space around you and convince yourself that whatever you looking at is not the present but the past. Since light takes time to reach from everything to reach you, everywhere you look, every sound you here is just an event in past. So wherever you point your fingertip, it's not a point in space or time but rather spacetime, you are pointing at an 'event'! And by doing our ongoing imaginations we are heading toward the future Our present is the sign we are here, the past is what we can see all around us and the future is what we are heading towards and we always will. What we are perceiving is the 4-dimensional spacetime. This spacetime is nothing but our universe itself. Let's study more about it from now onwards.

Forty-two

WHERE IS OUR UNIVERSE LOCATED

We are located in our universe which is completely made of a spacetime continuum. But did you ever wonder where it's located? Our universe is made of 4-dimensional spacetime, three of which are spatial and another one is the time dimension. What about extra dimensions? Yes! Our universe is located in the higher dimensions. We also call it hyperspace. We, humans, can just perceive up to 4 dimensions and not more than that, so hyperspace is not possible for us to imagine. The force of gravity even needs the extra dimensions to exist. Because gravity is the curvature in 4-dimensional spacetime and when you curve a particular dimension, you take it to the next dimension. Is there any other universe in hyperspace? Let's discuss that in the chapter ahead in this book. For now, let's note that the universe exists in the extra dimensions or hyperspace

Forty-three

THE OBSERVABLE UNIVERSE

A Notion Of A Term Distance or length Length is basically a physical entity which we all know what exactly is (roughly speaking), still to be clearer it's an interval in space. How long can a given length span up to? Try to imagine this! You probable might be infinitely long. But understand infinity is just our laziness to count. So this answer just does not make sense.. Setting boundaries for how large can a length be is equivalent to predicting how large our universe as a whole is. What we can define is the observable universe which is well-defined for us to imagine and mathematics to play its role. The length of the observable universe is basically our reach of eyesight to the farthest point in the universe at any given time. Our universe is 13.8 billion years old, thus the oldest light we can perceive is always to be the same as the age of our universe. How big is the observable universe? It is 93 billion light years across! The exaggerated (so that one can have an idea) image of our observable universe:

Our observable universe. The circular boundary is the cosmological horizon. Wait! if the age of the universe is 13.8 billion light years old then the farthest object we could be able to see or the length of the observable universe should be just 13.8 billion light years and not 93 billion light years away as that's the longest length that light can trace since 13.8 billion years and thus how can our observable universe be 93 billion light years across in length? Our universe according to Hubble's law our universe is expanding faster than the speed of light thus a milestone at 13.8 billion years has moved to 93 billion light years away or even much far. The light from these moving objects is still able to reach us as they get Redshifted or stretched and are not covering the extra distance stretched due to the cosmic expansion. More clearly (in this case) you can say that the light from 93 billion light years away reaches us within 13.8 billion years! Thus the Boundary of the observable universe which is also called the cosmological horizon lies somewhere 93 billion lightyears away from us. I know this chapter kind of got boring as you read ahead. I am mentioning this because I think that I still didn't give a sensible answer about how large our universe is. The reason behind the existence of this chapter in this book was just because I wanted to give a reader a viewpoint. So in next chapter lets fulfill this curiosity to know how large our universe exactly is.

Forty-four

HOW LARGE IS OUR UNIVERSE

Where does the universe end? or How large is our universe? We, humans, believe in discreteness, we believe that everything has a starting and an end, and we believe that everything has a limit. When this thought is applied to our universe, where does our universe end? Let's say for now I show you its end, then the next question would be what's beyond that end boundary? And so on. So don't you think our question about how large is our universe does not make any sense, that is because there is no sensible answer to the question. So let's now ask this question more intelligently. Our universe is made of spacetime and from nothing called 'space and time'. So our question which mentions just 'space' and not 'time' would be completely wrong. So since our universe is mad eof spacetime, wherever you point you just now point a point in space or an instant in time, you actually point an event. So every point in our universe represent an event. Our main question contained 'where' does your universe end or 'where' is its edge. So instead of thinking of the edge of the universe as a location in space, think of it as an event. We should instead ask where-when our universe will end. . So in that sense, our spacetime starts from the big bang and ends in a big crunch. So the boundaries of our universe actually is an 'event'.

But incase if you are impatient to wait until big crunch, you can witness the edge of universe right at the singularity of a black hole. Since the edge of the universe marks a break in spacetime which as we discussed are big bang and big crunch singularity. The same gateway is now we have in the form of a black hole, which also marks a break in the fabric of spacetime.

Let's see how can we understand our universe by one more perspective. For this, we need a bit of extra quantum physics theory.

Forty-five

THE WAVE AND PARTICLES

Quantum physics is the subject that predicts the behavior of sub-atomic particles, and which cant be described by any other field of physics. Quantum mechanics predicts the motion of particles with quite well accuracy if not approximately, this less precision or uncertainty in the subject of quantum mechanics, makes it a language of probability. According to quantum physics, every particle is a wave and vice versa. A light ray when passed through a slit, does show the interference pattern on the screen, showing the wavelike behavior. But when the same light when shun on a metal for the photoelectric effect, then light behaves like a particle. Not only light but this is seen to be happening with subatomic particles that do have a little mass as well. For example, in a nuclear reaction, an alpha particle does not possess much energy to penetrate through a potential barrier wall it possesses due to its electromagnetic interaction with rest of the subatomic particles in the radioactive atom, but the result we see is that it still makes it through its barrier. This phenomenon can only be explained by the wave behavior of the alpha particle. The alpha particle treated as a wave, and the probability for it to escape the potential barrier was always non-zero as a part of the wavefunction was always lies beyond the potential barrier. This makes alpha particle to transfer their existence to the other side of the barrier and again behave like a particle. The higher the speed of anything and the more massive it is, the less wavelike behavior anything posses. We call this a de Broglie wavelength which related the particle's wavelength with its mass and velocity. Since subatomic particles are very light in weight, they have long wavelengths associated with them. Normal-day objects are also made of subatomic particles, but they are connected via various fundamental forces of our universe. Treated every particle let's say of your couch the same way, they do possess significant wavelength at their scale. So that means that your couch is made of waves. But each of these waves is different from every other, spanning in either direction, making your couch as a whole become more difficult but not impossible to be treated as a wave. This applies to every normal day objects and larger ones.

Forty-six

THE OBSERVER AND THE UNIVERSE

Gravity is the force that makes gigantic bodies less wavelike and electromagnetic force is the force that makes normal day objects less wavelike. There is no problem for even a normal day object to behave like a wave. But they don't because of the irregularity in the wavefunctions of the atoms they are made of. So gravity and electromagnetic force are not allowing for the subatomic particles in any normal-day object or larger to behave wavelike entirely, if not individually. This makes the planets, stars, moon, oceans, etc behave like a particle. Without the particle nature of matter, life just wouldn't have existed. But the same wavelike nature of light heats our planet and makes alpha decay possible thing. The universe, its fundamental forces, the laws of physic the nature of matter, the fundamental constants, and many more parameters are well arranged at high precision in this universe to make life on our planet a possible thing. The universe needs observers to exist. Many of our experiments in quantum physics prove that the observation crystallizes the wavefunction or uncertainty into reality. This is the observer effect. The observation in the observer effect is just not confined by our direct or indirect measurement, but by our existence in this universe in general. So the observation that observers exist has reached up till the past until the big bang to crystallize the laws of physics that would make life a possible thing. Even a slight change of any of the fundamental constants say at a 10th decimal place is still enough to make life not a possible thing in this universe. The universe or spacetime since its formation from the big bang is very much connected to the observer now and in the future. Such a universe is called a participatory universe. And Our universe highly resembles it. See the image below:

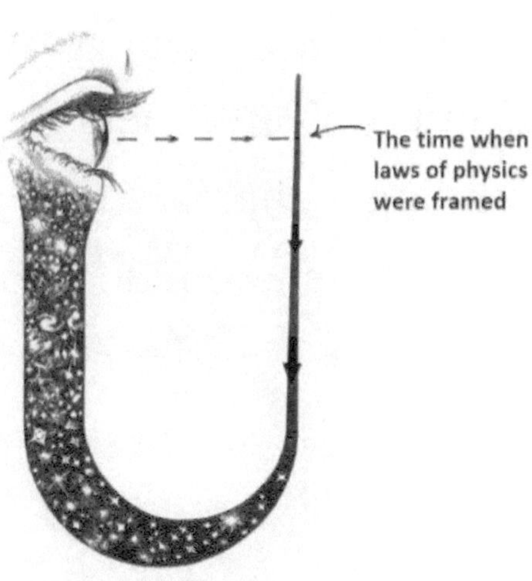

The above image is the description of the participatory universe and the shape of the diagram has nothing to do with the shape of our universe, it's just a short description of the entire content discussed above. The participatory universe says that the participants of this universe (we) have influenced this universe from its very start by making the observation now sitting on the earth. All fine, that we exist and make the universe to form in a way that would allow our existence. Literally, quantum physics has changed the way we ever looked at our universe. So, before we begin with the topic "The Fate of our universe" lets go with few more astounding topics on quantum mechanics. There Is even no problem If a reader direclty jumps over the next topic.

Forty-seven

THE SCHRODINGER'S CAT EXPERIMENT

Quantum mechanics or the mechanics going at the subatomic level is where all the parameters like position, energy, momentum, etc are not distinct (contrary to classical mechanics). Every particle has a wave associated with it. The smaller the mass the particle has, the larger its de Broglie wavelength would be (provided that the velocity of the particle is kept constant). Suppose we have a wall as a barrier for an electron to go to the other side, seeing the electron as a wave, in that case may allow the electron to get to the other side like a ghost! That is quantum tunneling. If a particle has its wavelength spread beyond its barrier then there are significant chances for a particle to tunnel the other side. In the nuclear decay process of radioactive atoms, the alpha particle needs to get past some coulomb barrier to get radiated. Because in terms of classical mechanics, the particle cannot escape the nuclei unless it's being provided some extra boost of energy. But, Quantum mechanics says that the alpha particle has a non-zero probability of escaping this barrier (by tunneling) and that is very true or otherwise alpha radiation would not be that common as it is. Schrodinger's cat experiment (thought experiment): Lets have a physicist we perform this experiment by having a cat , the nuclear reactor and the poison bottle as the apparatus in ant box . If the nuclear reactor reacts successfully as per the experiment by radiating an alpha particle (which is a matter of probability) then the nuclear energy is converted to the mechanical energy of a system which shatters a poison bottle which kills the cat inside the box. Let's say here we have a fifty-fifty percent chance of the nuclear reactor reacting successfully. If a nuclear reactor reacts then it converts its energy into mechanical energy which is associated with some setup to shatter the poison flask present in the box from which the cat inside the box will die and if it fails to react then the cat remains alive. So here in this experiment, the very quantum mechanical phenomenon of tunneling is translated from the subatomic (microscopic) to a macroscopic scale as the fate of a cat. Let's start the experiment by having the box closed. If I ask the physicist about the status of the cat inside the box then he will tell me a cat is dead as well as alive ! . To shortly divert from the topic , When we express say a position of a quantum particle by a wave then that outcome of probability is not in terms of 'or', it is all at a single instant result, thus it can be stated to be acquiring multiple positions at a single instant in time, it can also be said that a particle is in the superposition state. Returning back to the experiment I would say that the alpha particle has been radiated as well as it's not (the event is in the superposition of radiated and not radiated), thus it can be said that a cat is dead and as well as alive at a same time before the observation. But of course, as soon as the phycisist open the box there would be just a single outcome . What leads to the collapse of this wavefunction? It's the observation. Observation collapses the wavefunction. Can we say a conscious observer instead? Yes! Is there a moon if no one is observing it? are some of the famous questions which come to mind if one gets into this theory more deeply. Let's say a guy also famously known as a wigner's friend is observing (through a window) the physicist who was conducting the above experiment, so the status (imagine room as a box) of the physicist (like cat by physicist) is what is determined by the wigners friend as an observer and thus there lets consider a one more observer who is observing wigners friend (let's say) who is still in another room, deciding his status and so on. This demands a universal consciousness or a universal observer who collapse the wavefunction of status of each every inhabitant of this universe . Many relate this cosmic observer to be The God. Well! we don't know whether it's a god or anyone or any consciousness which is universal out there. All the thing to say is that this theory raises such questions which somehow may convince us regarding the existence of the universal consciousness. God may also exist in extra dimensions say in 10 or 11 dimensions. But I guess still the existence of god remains the question unless and until humanity has solid proof or evidence of it.

Forty-eight

THE QUANTUM ZENO PARADOX

Quantum Mechanics talk In a language of probability. In quantum mechanics, events in the future are described by the probability wavefunction. And the final outcome is what observation or measurement decides. Does observation refer to human observation particularly? It can be, but not necessarily as it can be even by the indirect human observation like detectors. Let's say we have an event where a whole event is described by some sort of wave function. That event be an electron shooting out of an electron gun to the metal plate. And let's say (In this case) we have only two results say an electron at initial position A and a final one over the metal plate let's call it B. Quantum mechanics describes this event in terms of probability. And lets say here we have a 50-50 chance of an electron being at state A and at state B. Let's say we observe an electron at state A or simply the wavefunction collapsed to point A. Then what after that? This is what we see in the normal world, we look at the things and we find one result. Now to rewrite that event of hitting a target A to an electron hitting point B instead, the wavefunction has to spread again so that choices (Point A or Point B) can again be made for an electron to land at point B. But what if we do not allow the wave function to spread out by continuously observing it? I mean not giving a mere chance for wavefunction of an electron to evolve from one discrete result so that it can jump to the other (point B) . This will halt the electron's motion at one particular point in space. And the motion will not evolve at all. This is the quantum Zeno effect. Which says that unaltered observation or measurement of the system halts its progress. So let's say we have an electron wave moving towards its target A and B. And before it hits one of the targets it was in the form of a wave. And when it hits the target it is in the form of a particle, as measurement is made. So well, before when its in the particle state it was a wave and It's the measurement that we make to collapse this wavefunction. Now quantum zeno effect says that on this journey of an electron wave hitting one of the two targets A and B, If we observe the electron before It hits the target, its wave function collapses right then and again when we don't look at it, then it redefines as a wave and If we keep doing this repeatedly, we actually then hinder Its motion to target A and B.

THE LAWS OF THERMODY -NAMICS

The laws of thermodynamics are very basic laws of nature. That's because they rely on symmetry, and this universe wants symmetry. Everything in this universe tends to attend the low energy state and tend to do that. To date, nothing we have ever seen has violated these laws. There are 4 laws in thermodynamics in total, lets's understand their meaning to understand the fate of our universe further. The first law of thermodynamics :

'IF TWO THERMODYNAMIC BODIES ARE EACH IN THE THERMODYNAMIC EQUILIBRIUM WITH THE THIRD BODY THEN ALL THREE BODIES ARE IN EQUILIBRIUM WITH EACH OTHER.'

This law is pretty obvious as it gives the idea of normal world heat transformation. And yes this law is what applies to every system of this universe. The transfer of heat is the process of becoming cold and to become cold is achieving the low energy state, which is what symmetry likes.

The second law of thermodynamics :

'THE TOTAL ENERGY IN OUR UNIVERSE REMAINS THE CONSTANT' The birth of the universe is the birth of energy. As our universe is finite, the energy in it is also finite. And every real thing in this universe is made out of this energy. The sun gives this energy to the earth, the plants on earth absorb this energy, herbivore feed on these plants so on. So this energy can be transformed from one form to the other, but never be created. This energy also includes quantum energy which lies everywhere in the universe at the Planck scale. This energy is stored in various forms. The form that can keep huge energy more than any other form is mass. In fact, mass is not a form of energy, it is infact same as energy. So in general we restate the second law of thermodynamics as follows :

'THE TOTAL MASS-ENERGY IN OUR UNIVERSE REMAINS THE CONSTANT'

Which is the very basic law in physics.

The third law of thermodynamics:

'THE ENTROPY OF THE UNIVERSE ALWAYS INCREASES

Entropy is the measure of the dispersal of energy through any process. It's a parameter we always take into account for any engineering project. Let's say an airplane is boarding at some speed. The fuel of an airplane just is not making the airplane move, but the airplane is making a lot of sounds. This decreases the efficiency of fuel. The sound of the airplane is somehow thus contributing to the entropy created by the airplane engine. Entropy is the 'information'. Our universe is the stock of events and everything, every point of the universe which Is made of spacetime represents an event, these events are contributing continuously to the total entropy of the universe. This second law of thermodynamics tells us that everything in this universe keeps on creating information. Aging is the very basic form of information creation that contributes to the total entropy of the universe. So everything in this universe does age. So entropy has its meaning attached to time. So until It moves in the future direction, the entropy of the universe will keep on increasing.

Third law of thermodynamics:

'THE ABSOLUTE ZERO TEMPERATURE IS UNATTAINABLE'

Yes! There is a limit to how much a system can be cooled down. An absolute zero kelvin or - 273.15 °C is that limit. Temperature is the measure of the movement of gas. In other words, this law states that a complete 100% percent vacuum is not possible. The entire universe is covered by quantum foam, so there is always some energy available, even in the empty space.

Fifty

THE FATE OF OUR UNIVERSE

We, humans, have made the universe attain the exact laws of physics that are necessary to allow our existence by our observation as the observers. But what if the entire universe chooses to end itself? Yes! the laws of thermodynamics give us the hint universe is going to end sometime in the future But what we humans can do to still ensure our future in the cosmic eternity when the entire universe decides to perish itself? The first law of thermodynamics says that the total energy in this universe remains constant. This energy is stored in various forms in nature. The very basic form of fuel we use is the food we eat, which is what we directly or indirectly get from plants, we get proteins from them, which are then converted into energy. The source of this energy is the sun as the plants use solar energy to prepare their food via photosynthesis. Later after that, we started using energy stored in the form of coal. It takes lakhs of years for the remains of animals and plants to form coal. We are still using the coal as our energy need in many sectors of industry. The energy in the coal has come from the energy of plants or animals that coal has formed from in fact have their energy is gained through plants which gain their energy from the sunlight. Due to science and technology, we are at a point where we can convert mass into energy and use that energy to do some work. Almost a century ago we were using all the forms of energy that the sun makes available to us. But later after then we discovered that mass is the same as energy, and we learned how to use the mass to run our turbines and make destructive bombs that would threaten humanity. The energy we get from the sun is even created from the mass-energy inside the sun itself. So all the energy available to us is in the form of solar energy which we have been using for decades and even now has the mass of the sun as its primary source. It's the energy required for life to evolve and has been achieved from the mass of the sun itself.

But, this story is not going to go on forever. The second law of thermodynamics says that entropy of the universe always increases. This entropy is the reason why everything ages and ultimately going to die out. Our sun continuously fusion its lighter elements into heavier ones. This fusion reaction will go on until we reach the iron. Iron being not able to fuse, our sun will Simply form a white dwarf which will again convert itself into a black dwarf and nothing else. Sun almost covers every energy need of us. And since it's not permanent as everything else in this universe, how are we going to shine our lamps? We can do that by harnessing the energy from another star system. But there are still 5 billion years more left for our sun to end all of Its fuel. But again the problem is that, as the sun fuses into the heavier elements, these elements are then sent into the outer layers of the sun. So our sun is continuously increasing its volume. So we just can't chill by looking at the sun to form a black dwarf like we watch television. The sun soon (next 4 to five billion years later) is going to come and fall on us. I mean the sun is going to get enough big as it expands that its atmosphere will completely swallow the earth. And as the sun is coming closer to us, more would be an increase in heat energy, so we have to leave earth much earlier. So we have a limited time to search for our next energy source In the next 2 billion years, humans would have developed enough technology that we can convert a mass of an asteroid completely into energy (say). But what is more convenient than a ready-made energy form? So, we definitely are going to look forward to visiting and getting settled around some other star first. Yes! The same star that one can see in the night sky. After harnessing the energy from around every star in this universe, we would look for harnessing the energy from the entire galaxy. After harnessing energy from every star, every galaxy, we would now seek energy from the last candidates which can stay alive till the cosmic eternity. The black holes would be the last energy source available source in this universe. They can provide us with an immense amount of energy. So in eternity, the black hole would be our last energy source. I guess we have gone too ahead in the future, we not even thought about, that the universe is ever going to end. The big crunch is the suspected end of our universe. So in eternity, let's see we get to know that our universe is shrinking its size by seeing the recession of the galaxies at the cosmic scale. Then what? We have to hurry up! We have to somehow again perform the tradition of securing our existence. But now what's next when the entire universe is dying? Ok, then let's travel to some other universe! Using the wormholes. The technology in eternity would be built enough for us to create a 5-dimensional bridge between the four-dimensional universes. And even we can find some readymade black holes serving this purpose for us. But wait! traveling to some other universe is just not as easy as one might think. As we need the laws of physics, including every fundamental constant over there to be the same as in our universe with a high precision. And provided we could find one, we going to do the entire mentioned process again and so on. So, we (the observers) are always going to secure our existence whatever happens till the eternity and in some other universe.

(SOME INTERESTING STUFF)

Fifty-one

FROM SPACETIME TO TIMESPACE

The spacetime which our universe is made of consists of three dimensional space and one dimensional space. When this spacetime gets into a black hole the space and time of spacetime switch there roles. The space becomes timelike and the time becomes spacelike. This switched role of spacetime is what I personally call "timespace" although this term may not exist in context of Astrophysics. It just helps us to encapsulate the whole imagination into a word. So lets call it the "timespace". Since now time has becomes spacelike, this doesn't mean that now we should give now three numbers for time. Remember, the only difference between space and time is that the time is a moving space. In infinite dimensional hyperspace the universe goes for first 4 of the spatial dimensions. The fourth of these spacetial dimension tend to flow with respect to the surrounding hyperspace. And since the lower three dimensions are connected to the time, they are carried with the flow itself. This recipe of spatial dimensions makes the universe.

So even when one sees the space and time switching there roles, its again just the same thing. The only thing which might dishtinguish the timespace and spacetime is may be the preserved imprints of the past of the spacetime in the timespace universe. So may be one be able to see the entire past of the spacetime before. This happens because the time dimension which we were locked in (providing no access to past and future), now curls and displays the entire past of itself. Not the future (addressed in the next chapter). And since now as an end result it still has three dimensions of time-switched space and other space-switched time. It works just a same way like spacetime.

Fifty-two

THE LIGHT'S PERSPECTIVE

The faster we move, the slower the time runs for us. Since light travels at the speed of causality, it almost stops elapsing for it. Imagine an observer looking at the sun, the photons from the sun in his perspective take 8 minutes to reach his eyes. But from the photon's perspective,e it made it into the observer's eye since the time it was emitted (since time doesn't elapse in the photon's perspective). So the question is is it predestined for a photon to ever go into the observer's eye? The answer is yes, but only predestined in the photon's perspective! Let's learn why. The below image shows the lightcone of a particle having some mass and in the inertial frame (at rest)

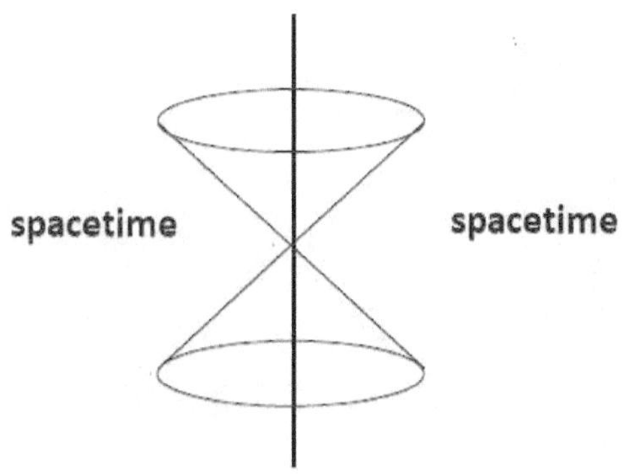

As the particle accelerates its lightcone gets tilted (remember, as we discussed). Let's say the lightcone tilts in the counterclockwise direction. The only permissible tilt any particle can go is the tilt of 45 degrees as after that tilting of lightcone is the same as moving faster than the speed of causality which is against the laws of physics. Light travels at the speed of causality, below image shows its lightcone.

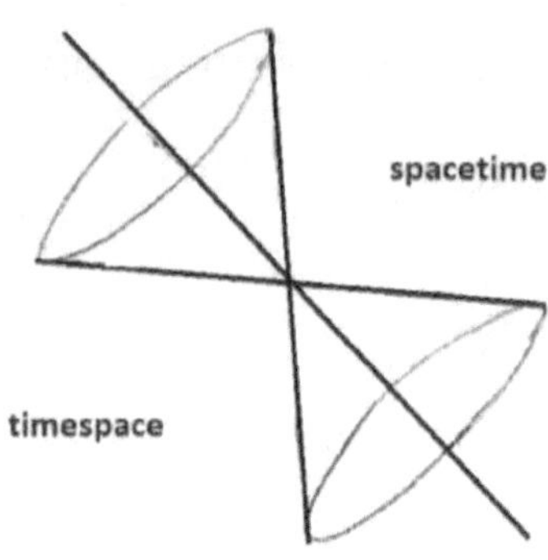

Light or photon, traveling at the speed of light always has its lightcone being 45 degrees tilted. And when spacetime gets tilted more than 45 degrees, it switches time to space and space to time in our Penrose diagram on which we define the light cone. In the above image of the lightcone of a photon, one can see that 50% part of the future and past lightcone enters the timespace and the other stays in spacetime. The term timespace is what I personally use to describe this new universe (one may not find the definition of this term on the web). As we live in spacetime, we can just see the spacetime behavior of light. But by studying the timespace + spacetime nature of light, many characteristics of light can be explained. For example why light only moves in one dimension? That's because, in time-space, space becomes timelike, so there is only one direction to move, while light moves in one dimension it can move in any direction as we see it, this is just due to the spacetime freedom of light.

Light or photons sits right in the middle of the timespace and spacetime universe. When one inhibits the timespace perspective he can see the entire timeline of the spacetime he was In before, curling itself into a three dimensional temporal reality combined with one dimensional space which makes the events in the universe past-present-future to be witnessed at a same time. Well. past and present is ok but the conjecture we just made may violate the quantum mechanical assertion that says that time is multi-fingured. So when you look at the future of spacetime in three-dimensional temporal reality of timespace, you should see multiple events of future happening at the same time.The formation of light marks the solid boundary in spaetime once its emitted and the event of observing that light in sense of multi figured nature of time is in hands of observer. For example, he can create a rindler horizon around himself and skip the solid boundary of the light. So if time is not multi-fingured and we have a single arrow of time, then it can be called that for few observers it was predestined to enter into their eye. And so if time is multifingured, the light only marks the solid boundary in spacetime, which can again be dodged by a worldline of an observer as a part of multi-fingured time. The statement that light marks the solid boundary in spacetime in both the cases (multi-fingured and not multi-fingured time) appies true. So the correct lies behind the question whether the time is multi-fingered or not. In my opinion, I believe in quantum mechanical assertion of multi-fingured nature of time.

Fifty-three

LIGHT AND ITS SPEED

What makes light so special that nothing can travel at its speed? Ok, to be honest, there is nothing special in light. The specialty is in the speed it travels at. From the start, we have been calling it the speed of light. But that speed limit has nothing to do with light. The actual value of this speed limit is 299 792 458 m / s. We call this the speed of causality. The speed of causality marks the boundary between causal relations or cause and effect. For example, in our light case , any cause that happened in past can affect us unless it lies inside our past lightcone. And similarly, any effect can't have any cause for us until it lies inside our future lightcone. The boundaries of the lightcone are margined by nothing but a speed limit. A speed limit that nothing in this universe can cross. This speed limit is what we call the speed of causality. The speed of causality is the universal speed limit of any given Lorentz frame. When anything tries to reach this speed, the more massive it becomes. And more is the energy we require to push anything to the speed of light. Since light and graviton are massless particles, they always travel at this speed.

Fifty-four

ELEMENTARY PARTICLE PHYSICS

(Introduction)

Having an atom into existence is not just the game of an electron, proton, and neutron. Proton and neutrons are made of 3 quarks, to hold or glue this quark one needs a gluon particle (strong force carrier). For an electron to maintain its energy level or say to interact with the proton in the atom itself needs virtual photons (electromagnetic force carrier), when an electron combines with a proton a neutron and a neutrino particle is produced. Again these particles have their own anti-particles. To give the fundamental particles like electrons, quarks, etc their masses we have Higgs bosons, which are also called a God Particle. These are just a few of the large sets of particles in existence. In reality, there can be innumerable particles in this universe, some of them are very mystical like the dark matter as we don't what exactly are made up of. The standard model of particle physics is all about it. How these particles interact with each other , etc. According to QFT (quantum field theory), the fundamental particles are nothing else but excitation in the quantum field. The Higgs field is a specific type of quantum field which gives the fundamental particles, their masses. The discovery of the Higgs boson was one of the greatest achievements of all time. Physical detection of such particles crystalizes the prediction made by the mathematics.

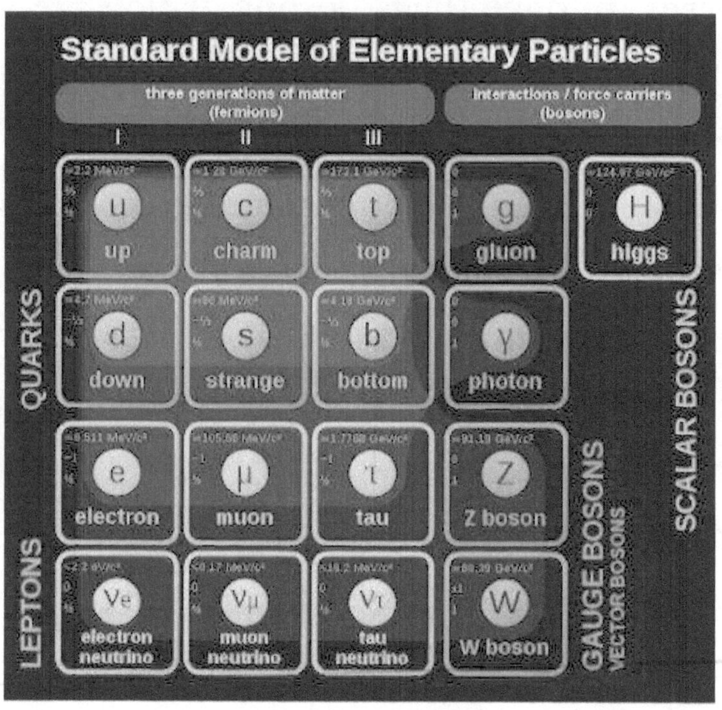

The standard model of particle physics Just after Bigbang, the Higgs field was inactive and thus every particle in that epoch were massless like photons and gravitons . Having no mass every particle there travelled at c (the speed of light). The large Hydron collider (LHC) is the particle accelerator which is located in Geneva, Switzerland. In LHC two or more particles are collided to get a stream of new particles being released at the instant of collision. One of the best experiment detection which makes up the holy grail of physics is the experimental discovery of graviton. Feynman Diagram by Richard Feynman which discusses various particle interactions in form of particle interaction diagrams for example:

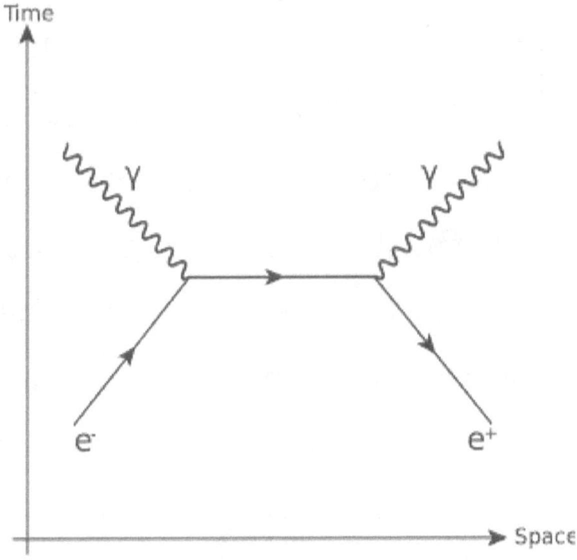

Here in this diagram, an electron pointing its temporal trajectory in the future interacts with a photon and instantaneously turns itself into a positron . If the current in one direction due to 176 the negative charge of an electron is taken as positive can be considered as having a negative value if we had a positron instead of an electron as the charge is flipped (mathematically one can check). Thus when an electron flow through spacetime gets deflected if changes its direction then can logically be considered as the flow of positron instead. That's so interesting! an electron traveling backward in time is a positron ! To consider any particle there can be infinite Feynman diagrams associated with it. Including the standard model, QFT, and Feymann diagram there are many theories and mathematical ways to describe the behaviour of subatomic and elementary particles. To conclude, particle physics gives us a lot of information about the time just after the big bang, also it's proving a great way to formulate a unified theory and many more.

Fifty-five

THE FORCE OF GRAVITY AND OTHER FUNDAMENTAL FORCES.

When we say force, it reminds us of an acceleration brought in mass. But we don't always have to push that mass by ourselves. In the case of the fundamental forces of the universe, this force seems very mystical. I broke the mystery of gravity by explaining to the reader what exactly it is and how it works. Every fundamental force in this universe needs a field to operate. The field responsible for weak force is the weak field, for the electromagnetic force, It's the electromagnetic field, and for strong force, it's the gluon field. Force the force of gravity the field is 4- dimensional spacetime itself! So the force of gravity is a very basic thing to have. Since the 190 fabric of spacetime is the fabric of reality, the fields of all fundamental forces have their meaning attached to the field that is responsible for the force of gravity. This is how the force of gravity Is different from other fundamental forces o this universe.

Fifty-six

THE MULTIVERSE HYPOTHESIS AND THE MANY WORLDS INTERPRETATION

In infinity war, Thanos shifted a huge part of an entire population to some other universe. But does a multiverse (which consists of many universes like and unlike us) exist? The fact that our universe is not alone in the hyperspace (extra dimensions) is the Multiverse Hypothesis. Thus we have upgraded our union set of universes to the multiverse. The many worlds interpretation: Quantum mechanics is the language of probability. It assigns every particle,, not a discrete existence but a waveform. This wave is what the particle is and vice versa. The low and high peaks of the wave including together represent existing universes. So our universe is just one of those peaks. It would be wrong to imagine the wavefunction of the multiverse as a 2 or 3-dimensional graph, as one should remember that we are talking about the wavefunction in Hyperspace! , The reality which we perceive is due to our mere observation which collapses the wave function to produce some discrete results. One might question then that what about the other possibilities ? where do they go even if they were not made discrete by our observation in our universe. The collapse of function is not only dependent on the consciousness and the observation of the observer but also on the environmental conditions around it. Thus the events which had no discreteness in our universe are what are at the peaks of some other universe. So maybe a Schrodinger cat is alive in our universe, but the dead part of it is somewhere in another universe. Thus to comment on the wave function, no part of it is unreal and it's our observation and the conditions around which make the particular part of a wavefunction to peak and achieve discreteness or to collapse in ours, if not in another universe. In short wave function is a combination of the various outcomes in the various universes in the multiverse.

Fifty-seven

WHY GALAXIES LIKE OURS (MILKY WAY) ARE FLAT, WHILE THE PLANETS AND STARS SPHERICAL IN SHAPE

If you are in hurry then just read the summarization mentioned down and that's it (sufficient to answer the main topic of this article) Although reading the main content has its own taste and deep understanding. There are galaxies with a variety of shapes in our universe. This article is limited to only spiral galaxies (eg our Milkyway galaxy). The laws of physics are the same for each and every constituent or inhabitant of this universe. Dating back to several billion years in the past (even before when galaxies were forming in our universe). That was the time when individual particles were interacting with each other gravitationally to form some cloud-like structure which is our future galaxy and still 194 further processing will get our planets to us. But what mechanism gives our galaxy the disk structure it has (the spiral galaxy), while most of its inhabitants like planets, stars, and also some asteroids are spherical in shape. The underlining force which clumps the matter into galaxies or planets is basically the same (gravity). Then why is there a difference in structure? In a nutshell, it's just due to the way and condition of formation. The Formation Of Galaxies : The formation of the spiral galaxies is a much slower process as compared to the planets and stars. In the case of galaxies, individual particles (separated by wide distances) interact with each other gravitationally, until it forms a rotating supermassive black hole. Once the rotating supermassive black hole forms, the chain continues further. The centrifugal force due to the revolution of particles around the axis of rotation of the SMBH (supermassive black hole) maintains the matter in the vicinity uninteracted. This structure of revolution of particles around the rotating SMBH further gets a disk-like appearance. The influence of matter above the disk on what is inside the disk nulls off due to the interaction of matter which is below the disk (equal and opposite forces). The only net forces now remain is the gravitational influence of the matter inside the disk on the matter above and beneath the plane (downwards). Thus the net force on the matter below and above the disk remains downward towards the planet of revolution. This overall process gives a system a disk-like look. Refer to the following image:

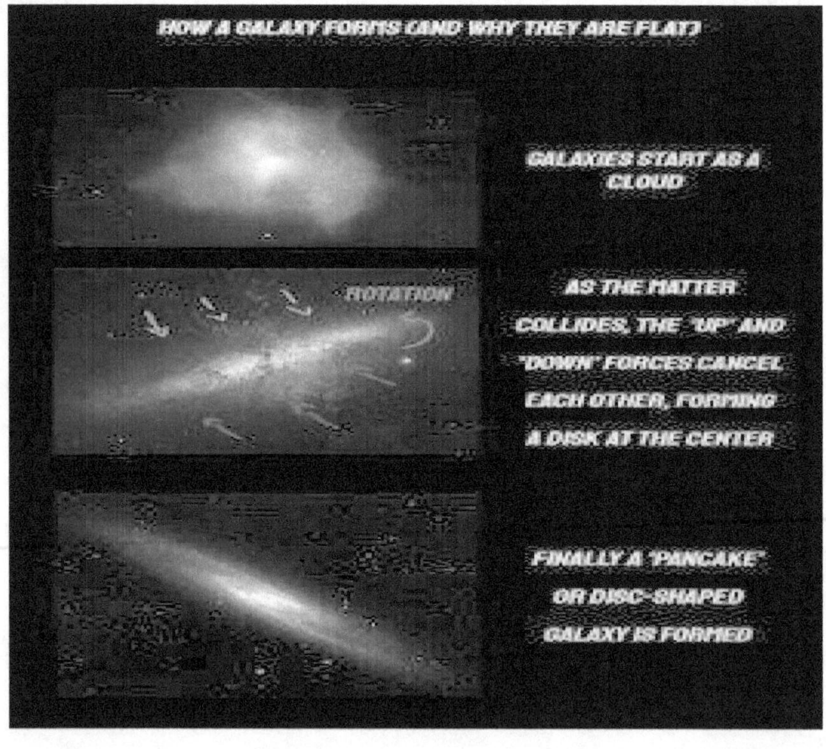

The Formation Of The Planets : In the case of planet formation mostly what collides are the bulk particles and rarely finer ones. Again as the matter collides it gains some angular momentum to itself and individual bulk particles say above and below the plane of rotation approach the system and combine with it. In the case of the formation of galaxies the particles were finer and were assumed by us to be symmetrically oriented although widely separated (a very close situation to what is in reality) which makes our force cancellation statement a foundation to result. In the case of planets, the particles are bulkier. Thus the probability of symmetric orientation among particles is pretty less, so the force doesn't cancel out over here and and the bulkier particles combine to form a spherical planet .

The Formation Of The Stars : In the case of a star, the matter interacting although is small in size (maybe of what is in the case of galaxies), the difference that appears is the formation of SMBH . In case of star formation SMBH do not form . Thus no accretion disk . And what remains is just symmetric attraction of particles equal from all side , which gives star a structure of a perfect sphere. And once the fusion reaction starts we have our prototype star. 198 The Summarization : In the case of planet formation, bulkier particles are unsymmetrically distributed and thus there is a net inward force on every chunk out there. And as the gravity is the same from all sides the plane takes a spherical shape. In the case of stars, the particles are finer and less widely separated and also are symmetrically distributed. And as there is no formation of the violent disk due to absence of SMBH . All matter is attracted symmetrically In the case of galaxies, the atoms are finer and very widely separated from each other. Also, the particles during formation are symmetric only at large scales. Any rare event of finding a concentration of these particles at a particular point (say) makes that point the center of the galaxy. As time passes due to the constant clumping of matter, a rotating supermassive black hole forms which result in the formation of an accretion disk (disk in which highly dense matter revolve around the black hole) around it. The force of gravity due to matter above and beneath the accretion disk cancels out due to equal and opposite directions of the force. Thus the net force of the matter above and below the accretion disk is towards the disk due to the gravitational influence of the accretion disk . And thus the whole stuff becomes the part of that accretion disk that we call our galaxy. Thus even if the underlining force behind behind the formation of stars, planets and galaxies is the same, the nature of the matter , its position in space, distance between individual particles are what decide the shape of the structure formed.

Fifty-eight

ARE WE ALONE IN THIS UNIVERSE

To this date, there is no solid evidence (Although many claim there are) of the existence of aliens. But still, are there any one of them out there? Any planet in another system in the habitable zone (zone suitable for the creation of life) gives us the chance of having extraterrestrial life. Our planet earth, which is in the solar system, is in the miky way that has more than 3200 star systems in it and still there are more than three trillion estimated galaxies in the observable universe. Thus in that sense, it's highly probable for the alien lives to exist. Can Extraterrestrial life if exist be more intelligent than us? well! we don't know. Aliens may even be type 2 civilizations at the level of which technology is enough advanced to harness the complete power of any star. Aliens also may be advanced human beings. If they meet us we would have a great opportunity to learn from them. Also, they can be bacteria or any unicellular animal too.

Jupiter's moon Europa is quite likely to sustain life in it as there remains the liquid water underneath its ice sheets which have widths of several kilometers wide. Europa gets sunlight but not that bright to maintain the water in a liquid state beneath So from where does Europa get that energy to maintain the liquid water beneath the ice sheets? the answer is the tidal gravity of Jupiter. The tidal gravity or tidal force of Jupiter brings out some deformation in the shape of Europa, this mechanical energy is then 210 converted into heat energy by breaking various molecular bonds providing some heat energy source to maintain the liquid water under the ice sheets of Europa. One also can compare this with the tennis ball when it gets hitted by a racket. When tennis ball is hitted by a tennis racket, it gets deformed and infact get heated due to the mechanical deformation acted on it. In the same way Europa receives it's heat by which it maintains liquid water beneath the icesheets on it. But here what is deforming Europa is the tidal gravity of Jupiter. This deformation is not as great as in tennis ball analogy but quite enough to provide energy to the planet to keep water in the liquid form beneath the ice sheets on it. A mission is planned by NASA till 2024 to study still in detail to investigate the suitable conditions for the existence of life on Europa and if we get some positive results there, then it would truely be one of the greatest achievements of us. Well! what about the aliens from some other universe? Great idea right! If the multiverse exists, then we have a greater chance of life harboring in each of the individual universes in it . So if we are alone in 'our universe ' then there might be aliens existing in another universe too. What about the bulk beings? (Multidimensional beings) There is no proof of the existence of bulk beings. If God exists in 10 or 11 dimensions we may call him a bulk being or bulk aliens or whatever you like. There is a famous equation of aliens which gives the probability of the existence of extraterrestrial life in our universe. It considers each and every parameter like the number of planets which are habitable, the fraction of stars orbited by planets, etc. This equation is also known as

$$N = R_* \cdot f_p \cdot n_e \cdot f_l \cdot f_i \cdot f_c \cdot L$$

N = number of civilizations with which humans could communicate

R_* = mean rate of star formation

f_p = fraction of stars that have planets

n_e = mean number of planets that could support life per star with planets

f_l = fraction of life-supporting planets that develop life

f_i = fraction of planets with life where life develops intelligence

f_c = fraction of intelligent civilizations that develop communication

L = mean length of time that civilizations can communicate

The presence of us in this universe as the only life in existence in actuality is a rare case. If we are in reality all alone in our universe, could it be interpreted as there being some relation 212 between the existence of humanity and the existence of the universe? Roughly speaking it could be! Even some observations of cosmic microwave background like the axis of evil suggest us that we humans have some special role in this universe! But still, this is speculation and there is no solid basis proving it strongly. But until we investigate each and every suspected habitable zone for life, our hunger of searching for extraterrestrial life will persist.